ANTI-VIRUS TOOLS AND TECHNIQUES
FOR COMPUTER SYSTEMS

ANTI-VIRUS TOOLS
AND TECHNIQUES
FOR COMPUTER SYSTEMS

W. Timothy Polk   **Lawrence E. Bassham III**
John P. Wack   **Lisa J. Carnahan**

National Institute of Standards and Technology

Advanced Computing
and
Telecommunications Series

NOYES DATA CORPORATION
Park Ridge, New Jersey, U.S.A.

Copyright © 1995 by Noyes Data Corporation
Library of Congress Catalog Card Number: 94-31166
ISBN: 0-8155-1364-X
Printed in the United States

Transferred to Digital Printing, 2011
Printed and bound in Great Britain by
CPI Antony Rowe, Chippenham and Eastbourne

Library of Congress Cataloging-in-Publication Data

Anti-Virus tools and techniques for computer systems / by W. Timothy
    Polk.
              p.   cm. -- (Advanced computing and telecommunications series)
        Includes bibliographical references (p.   ).
        ISBN 0-8155-1364-X
        1. Computer viruses. I. Polk, William T. II. Series.
QA76.76.C68A68   1995
005.8--dc20   94-31166
                                                   CIP

Preface

Computer viruses continue to pose a threat to the integrity and availability of computer systems. This is especially true for users of personal computers. A variety of anti-virus tools are now available to help manage this threat. These tools use a wide range of techniques to detect, identify, and remove viruses.

Part I provides criteria for judging the functionality, practicality, and convenience of anti-virus tools. It furnishes information which readers can use to determine which tools are best suited to target environments, but it does not weigh the merits of specific tools. Part I is based on "Guide to Selection of Anti Virus Tools and Techniques," published by NIST (800-5).

Part II contains guidance for managing the threats of computer viruses and related software and unauthorized use. It is geared towards managers of end-user groups, managers dealing with multi-user systems, personal computers and networks. The guidance is general and addresses the vulnerabilities that are most likely to be exploited. It emphasizes that organizations cannot effectively reduce their vulnerabilities to viruses and related threats unless the organization commits to a virus prevention program, involving the mutual cooperation of all computer managers and users. The guidance is aimed at helping managers prevent and deter virus attacks, detect when they occur or are likely to occur, and then to contain and recover from any damage caused by the attack.

The virus prevention program centers on strong user education, software management, the effective use of system controls, monitoring of user and system activity to detect abnormalities, and contingency procedures for containing and recovering. It contains an overview of viruses and related software, and several chapters of guidance for managers of multiuser computers, managers and users of personal computers, managers of wide and local area networks including personal computer networks, and managers of end-user groups. A reading list of supplementary documentation is provided. Part II is based on "Computer Viruses and Related Threats: A Management Guide," published by NIST (500-166).

This book is intended primarily for technical personnel selecting anti-virus tools for an organization. Additionally, this book is useful for personal computer end-users who wish to select appropriate solutions for their own system.

Notice

The materials in this book were prepared as accounts of work sponsored by a government agency. On this basis the Publisher assumes no responsibility nor liability for errors or any consequences arising from the use of the information contained herein.

The book is sold with the understanding that the Publisher is not engaged in rendering legal, technical, or other professional service. If advice or other expert assistance is required, the service of a competent professional should be sought.

Contents and Subject Index

PART II
COMPUTER VIRUSES AND RELATED THREATS:
A MANAGEMENT GUIDE

Part I

Guide to the Selection
of Anti-Virus
Tools and Techniques

1. Introduction

This document provides guidance in the selection of security tools for protection against computer viruses. The strengths and limitations of various classes of anti-virus tools are discussed, as well as suggestions of appropriate applications for these tools. The technical guidance in this document is intended to supplement the guidance found in NIST Special Publication 500-166, *Computer Viruses and Related Threats: A Management Guide* [WC89], included as Part II of this book.

This document concentrates on widely available tools and techniques as well as some emerging technologies. It provides general guidance for the selection of anti-virus tools, regardless of platform. However, some classes of tools, and most actual products, are only available for personal computers. Developers of anti-virus tools have focused on personal computers since these systems are currently at the greatest risk of infection.

1.1 Audience and Scope

This document is intended primarily for technical personnel selecting anti-virus tools for an organization. Additionally, this document is useful for personal computer end-users who wish to select appropriate solutions for their own system. This document begins with an overview of the types of functionality available in anti-virus products and follows with selection criteria which must be considered to ensure practicality and convenience. The body of the document describes specific classes of anti-virus tools (e.g., scanners) in terms of the selection criteria. This document closes with a summary comparing the different classes of tools and suggests possible applications.

The guidance presented in this document is general in nature. The document makes no attempt to address specific computer systems or anti-virus tools. However, at this time the computer virus problem is most pressing in the personal computer arena. Consequently, most types of anti-virus tools are available as personal computer products. As a result, some information will address that specific environment.

Certain commercial products are identified in this paper in order to adequately specify procedures being described. In no case does such identification imply recommendation or endorsement by the National Institute of Standards and Technology, nor does it imply that the material identified is necessarily the best for the purpose.

1.2   How to Use This Document

The remainder of this section is devoted to terminology and basic concepts.

Section 2 describes the different types of functionality that are available in anti-virus tools. Several different types of detection tools are described, as well as identification and removal tools. This information should assist readers in identifying the classes of products appropriate for their environment.

Section 3 describes some critical selection factors, including accuracy, ease of use, and efficiency. The description of each of these factors is dependent on the functional class of product in question. These selection factors are used to describe product classes in the sections that follow.

Section 4 describes specific classes of tools, such as scanners or checksum programs, and the techniques they employ. This section provides the reader with detailed information regarding the functionality, accuracy, ease of use and efficiency of these classes of tools.

Section 5 presents guidelines for the selection of the most appropriate class of anti-virus tools. It begins by outlining the important environmental aspects that should be considered. Next, the information from Section 4 is summarized and a variety of tables comparing and contrasting the various classes of tools are presented. The remainder of the section provides several hypothetical user scenarios. A battery of tools is suggested for each application.

Section 6 presents guidelines for the selection of the best tool from within a particular class. Important features that may distinguish products from others within a particular class are highlighted.

This document will be most useful if read in its entirety. However, the reader may wish to skip the details on different tools found in Section 4 on an initial reading. Section 5 may help the reader narrow the focus to specific classes of tools for a specific environment. Then the reader may return to Section 4 for details on those classes of tools.

1.3   Definitions and Basic Concepts

This section presents informal definitions and basic concepts that will be used throughout the document. This is intended to clarify the meaning of certain terms which are used inconsistently in the virus field. However, this section is not intended as a primer on viruses. Additional background information and an extensive "Suggested Reading" list may be found in both Part I and Part II.

A **virus** is a self-replicating code segment which must be attached to a host executable.[1] When the host is executed, the virus code also executes. If possible, the virus will replicate by attaching a copy of itself to another executable. The virus may include an additional "payload" that triggers when specific conditions are met. For example, some viruses display a message on a particular date.

A **Trojan horse** is a program that performs a desired task, but also includes unexpected (and undesirable) functions. In this respect, a Trojan horse is similar to a virus, except a Trojan horse does not replicate. An example of a Trojan horse would be an editing program for a multi-user system which has been modified to randomly delete one of the user's files each time that program is used. The program would perform its normal, expected function (editing), but the deletions are unexpected and undesired. A host program that has been infected by a virus is often described as a Trojan horse. However, for the purposes of this document, the term Trojan horse will exclude virus-infected programs.

A **worm** is a self-replicating program. It is self-contained and does not require a host program. The program creates the copy and causes it to execute; no user intervention is required. Worms commonly utilize network services to propagate to other computer systems.

A **variant** is a virus that is generated by modifying a known virus. Examples are modifications that add functionality or evade detection. The term variant is usually applied only when the modifications are minor in nature. An example would be changing the trigger date from Friday the 13th to Thursday the 12th.

An **overwriting** virus will destroy code or data in the host program by replacing it with the virus code. It should be noted that most viruses attempt to retain the original host program's code and functionality after infection because the virus is more likely to be detected and deleted if the program ceases to work. A **non-overwriting** virus is designed to append the virus code to the physical end of the program or to move the original code to another location.

A **self-recognition** procedure is a technique whereby a virus determines whether or not an executable is already infected. The procedure usually involves searching for a particular value at a known position in the executable. Self-recognition is required if the virus is to avoid multiple infections of a single executable. Multiple infections cause excessive growth in size of infected executables and corresponding excessive storage space, contributing to the detection of the virus.

A **resident** virus installs itself as part of the operating system upon execution of an infected host program. The virus will remain resident until the system is shut down. Once installed in memory, a resident virus is available to infect all suitable hosts that are accessed.

---

[1] An executable is an abstraction for programs, command files and other objects on a computer system that can be executed. On a DOS PC, for example, this would include batch command files, COM files, EXE-format files and boot sectors of disks.

A **stealth** virus is a resident virus that attempts to evade detection by concealing its presence in infected files. To achieve this, the virus intercepts system calls which examine the contents or attributes of infected files. The results of these calls must be altered to correspond to the file's original state. For example, a stealth virus might remove the virus code from an executable when it is read (rather than executed) so that an anti-virus software package will examine the original, uninfected host program.

An **encrypted** virus has two parts: a small decryptor and the encrypted virus body. When the virus is executed, the decryptor will execute first and decrypt the virus body. Then the virus body can execute, replicating or becoming resident. The virus body will include an encryptor to apply during replication. A **variably encrypted** virus will use different encryption keys or encryption algorithms. Encrypted viruses are more difficult to disassemble and study since the researcher must decrypt the code.

A **polymorphic** virus creates copies during replication that are functionally equivalent but have distinctly different byte streams. To achieve this, the virus may randomly insert superfluous instructions, interchange the order of independent instructions, or choose from a number of different encryption schemes. This variable quality makes the virus difficult to locate, identify, or remove.

A **research** virus is one that has been written, but has never been unleashed on the public. These include the samples that have been sent to researchers by virus writers. Viruses that have been seen outside the research community are termed **"in the wild."**

It is difficult to determine how many viruses exist. Polymorphic viruses and minor variants complicate the equation. Researchers often cannot agree whether two infected samples are infected with the same virus or different viruses. We will consider two viruses to be different if they could not have evolved from the same sample without a hardware error or human modification.

2. Functionality

Anti-virus tools perform three basic functions. Tools may be be used to *detect, identify,* or *remove* viruses.[2] Detection tools perform proactive detection, active detection, or reactive detection. That is, they detect a virus before it executes, during execution, or after execution. Identification and removal tools are more straightforward in their application; neither is of use until a virus has been detected.

2.1 Detection Tools

Detection tools detect the existence of a virus on a system. These tools perform detection at a variety of points in the system. The virus may be actively executing, residing in memory, or stored in executable code. The virus may be detected before execution, during execution, or after execution and replication.

2.1.1 Detection by Static Analysis

Static analysis detection tools examine executables without executing them. Such tools can be used in proactive or reactive fashion. They can be used to detect infected code before it is introduced to a system by testing all diskettes before installing software on a system. They can also be used in a more reactive fashion, testing a system on a regular basis to detect any viruses acquired between detection phases.

2.1.2 Detection by Interception

To propagate, a virus must infect other host programs. Some detection tools are intended to intercept attempts to perform such "illicit" activities. These tools halt the execution of virus-infected programs as the virus attempts to replicate or become resident. Note that the virus has been introduced to the system and attempts to replicate before detection can occur.

---

[2]A few tools are designed to *prevent* infection by one or more viruses. The discussion of these tools is limited to Section 4.7.2, *Inoculation*, due to their limited application.

2.1.3   Detection of Modification

All viruses cause modification of executables in their replication process. As a result, the presence of viruses can also be detected by searching for the unexpected modification of executables. This process is sometimes called *integrity checking*.

Detection of modification may also identify other security problems, such as the installation of Trojan horses. Note that this type of detection tool works only after infected executables have been introduced to the system *and the virus has replicated.*

2.2   Identification Tools

Identification tools are used to identify *which* virus has infected a particular executable. This allows the user to obtain additional information about the virus. This is a useful practice, since it may provide clues about other types of damage incurred and appropriate clean-up procedures.

2.3   Removal Tools

In many cases, once a virus has been detected it is found on numerous systems or in numerous executables on a single system. Recovery from original diskettes or clean backups can be a tedious process. Removal tools attempt to efficiently restore the system to its uninfected state by removing the virus code from the infected executable.

3. Selection Factors

Once the functional requirements have been determined, there will still be a large assortment of tools to choose from. There are several important selection factors that should be considered to ensure that the right tool is selected for a particular environment.

There are four critical selection factors: *Accuracy, Ease of Use, Administrative Overhead* and *System Overhead*. Accuracy describes the tool's relative success rate and the types of errors it can make. Ease of use describes the typical user's ability to install and execute the tool and interpret the results. Administrative overhead is the measure of technical support and distribution effort required. System overhead describes the tool's impact on system performance. These factors are introduced below. In depth discussions of these factors are in subsequent subsections.

Accuracy is the most important of the selection factors. Errors in detecting, identifying or removing viruses undermine user confidence in a tool, and often cause users to disregard virus warnings. Errors will at best result in loss of time; at worst they will result in damage to data and programs.

Ease of use is concerned with matching the background and abilities of the system's user to the appropriate software. This is also important since computer users vary greatly in technical skills and ability.

Administrative overhead can be very important as well. Distribution of updates can be a time-consuming task in a large organization. Certain tools require maintenance by the technical support staff rather than the end-user. End-users will require assistance to interpret results from some tools; this can place a large burden on an organization's support staff. It is important to choose tools that your organization has the resources to support.

System overhead is inconsequential from a strict security point of view. Accurate detection, identification or removal of the virus is the important point. However, most of these tools are intended for end-users. If a tool is slow or causes other applications to stop working, end-users will disable it. Thus, attention needs to be paid to the tool's ability to work quickly and to co-exist with other applications on the computer.

3.1 Accuracy

Accuracy is extremely important in the use of all anti-virus tools. Unfortunately, all anti-virus tools make errors. It is the type of errors and frequency with which they occur that is important. Different errors may be crucial in different user scenarios.

Computer users are distributed over a wide spectrum of system knowledge. For those users with the system knowledge to independently verify the information supplied by an anti-virus tool, accuracy is not as great a concern. Unfortunately, many computer users are not prepared for such actions. For such users, a virus infection is somewhat frightening and very confusing. If the anti-virus tool is supplying false information, this will make a bad situation worse. For these users, the overall error rate is most critical.

3.1.1   Detection Tools

Detection tools are expected to identify all executables on a system that have been infected by a virus. This task is complicated by the release of new viruses and the continuing invention of new infection techniques. As a result, the detection process can result in errors of two types: *false positives* and *false negatives*.

When a detection tool identifies an uninfected executable as host to a virus, this is known as a *false positive* (this is also known as a Type I error.) In such cases, a user will waste time and effort in unnecessary cleanup procedures. A user may replace the executable with the original only to find that the executable continues to be identified as infected. This will confuse the user and result in a loss of confidence in either the detection procedures or the tool vendor. If a user attempts to "disinfect" the executable, the removal program may abort without changing the executable or will irreparably damage the program by removing useful code. Either scenario results once more in confusion for the user and lost confidence.

When a detection tool examines an infected executable and incorrectly proclaims it to be free of viruses, this is known as a *false negative*, or Type II error. The detection tool has failed to alert the user to the problem. This kind of error leads to a false sense of security for the user and potential disaster.

3.1.2   Identification Tools

Identification tools identify *which* virus has infected a particular executable. Defining failure in this process turns out to be easier than success. The identification tool has failed if it cannot assign a name to the virus or assigns the wrong name to the virus.

Determining if a tool has correctly named a virus should be a simple task, but in fact it is not. There is disagreement even within the anti-virus research community as to what constitutes "different" viruses. As a result, the community has been unable to agree on the number of existing viruses, and the names attached to them have only vague significance. This leads to a question of *precision*.

As an example, consider two PC virus identification tools. The first tool considers the set of PC viruses as 350 distinct viruses. The second considers the same set to have 900 members. This occurs because the first tool groups a large number of variants under a single name. The second tool will name viruses with greater precision (i.e., viruses grouped together by the first tool are uniquely named by the second).

Such precision problems can occur even if the vendor attempts to name with high precision. A tool may misidentify a virus as another variant of that virus for a variety of reasons. The variant may be new, or analysis of samples may have been incomplete. The loss of precision occurs for different reasons, but the results are no different from the previous example. Any "successful" naming of a virus must be considered along with the degree of precision.

3.1.3   Removal Tools

Removal tools attempt to restore the infected executables to their uninfected state. Removal is successful if the executable, after disinfection, matches the executable before infection on a byte-for-byte basis. The removal process can also produce two types of failures: *hard failure* and *soft failure*.

A *hard failure* occurs if the disinfected program will no longer execute or the removal program terminates without removing the virus. Such a severe failure will be obvious to detect and can occur for a variety of reasons. Executables infected by overwriting viruses cannot be recovered in an automated fashion; too much information has been lost. Hard failures also occur if the removal program attempts to remove a different virus than the actual infector.

Removal results in a *soft failure* if the process produces an executable, which is slightly modified from its original form, that can still execute. This modified executable may never have any problems, but the user cannot be certain of that. The soft failure is more insidious, since it cannot be detected by the user without performing an integrity check.

3.2   Ease of Use

This factor focuses on the level of difficulty presented to the end-user in using the system with anti-virus tools installed. This is intended to gauge the difficulty for the system user to utilize and correctly interpret the feedback received from the tool. This also measures the increased difficulty (if any) in fulfilling the end-user's job requirements.

Ease of Use is the combination of utilization and interpretation of results. This is a function of tool design and quality of documentation. Some classes of tools are inherently more difficult to use. For example, installation of the hardware component of a tool requires greater knowledge of the current hardware configuration than a comparable software-only tool.

3.3   Administrative Overhead

This factor focuses on the difficulty of administration of anti-virus tools. It is intended to gauge the workload imposed upon the technical support team in an organization.

This factor considers difficulty of installation, update requirements, and support levels required by end-users. These functions are often the responsibility of technical support staff or system administrators rather than the end-user. Note that an end-user without technical support must perform all of these functions himself.

3.4   System Overhead

System overhead measures the overall impact of the tool upon system performance. The relevant factors will be the raw speed of the tool and the procedures required for effective use. That is, a program that is executed every week will have a lower overall impact than a program that runs in the background at all times.

4. Tools and Techniques

There is a wide variety of tools and techniques which can be applied to the anti-virus effort. This section will address the following anti-virus techniques:

- signature scanning and algorithmic detection
- general purpose monitors
- access control shells
- checksums for change detection
- knowledge-based removal tools
- research efforts
  - heuristic binary analysis
  - precise identification
- other tools
  - system utilities as removal tools
  - inoculation

For detection of viruses, there are five classes of techniques: signature scanning and algorithmic detection; general purpose monitors; access control shells; checksums for change detection; and heuristic binary analysis. For identification of viruses, there are two techniques: scanning and algorithmic detection; and precise identification tools. Finally, removal tools are addressed. Removal tools come in three forms: general system utilities, single-virus disinfectors, and general disinfecting programs.

4.1 Signature Scanning and Algorithmic Detection

A common class of anti-virus tools employs the complementary techniques of signature scanning and algorithmic detection. This class of tools is known as scanners, which are static analysis detection tools (i.e., they help detect the presence of a virus). Scanners also perform a more limited role as identification tools (i.e., they help determine the specific virus detected). They are primarily used to detect if an executable contains virus code, but they can also be used to detect resident viruses by scanning memory instead of executables.

They may be employed proactively or reactively. Proactive application of scanners is achieved by scanning all executables introduced to the system. Reactive application requires scanning the system at regular intervals (e.g., weekly or monthly).

4.1.1 Functionality

Scanners are limited intrinsically to the detection of known viruses. However, as a side effect of the basic technique, some new variants may also be detected. They are also identification tools, although the methodology is imprecise.

Scanners examine executables (e.g., .EXE or .COM files on a DOS system) for indications of infection by known viruses. Detection of a virus produces a warning message. The warning message will identify the executable and name the virus or virus family with which it is infected. Detection is usually performed by signature matching; special cases may be checked by algorithmic methods.

In signature scanning an executable is searched for selected binary code sequences, called a virus signature, which are unique to a particular virus, or a family of viruses. The virus signatures are generated by examining samples of the virus. Additionally, signature strings often contain wild cards to allow for maximum flexibility.

Single-point scanners add the concept of relative position to the virus signature. Here the code sequence is expected at a particular position within the file. It may not even be detected if the position is wrong. By combining relative position with the signature string, the chances of false positives is greatly reduced. As a result, these scanners can be more accurate than blind scanning without position.

Polymorphic viruses, such as those derived from the *MtE* (mutation engine) [Sku92], do not have fixed signatures. These viruses are self-modifying or variably encrypted. While some scanners use multiple signatures to describe possible infections by these viruses, algorithmic detection is a more powerful and more comprehensive approach for these difficult viruses.

4.1.2 Selection Factors

Accuracy

Scanners are very reliable for identifying infections of viruses that have been around for some time. The vendor has had sufficient time to select a good signature or develop a detection algorithm for these well-known viruses. For such viruses, a detection failure is unlikely with a scanner. An up-to-date scanner tool should detect and to some extent identify any virus you are likely to encounter. Scanners have other problems, though. In the detection process, both false positives and false negatives can occur.

False positives occur when an uninfected executable includes a byte string matching a virus signature in the scanner's database. Scanner developers test their signatures against libraries of commonly-used, uninfected software to reduce false positives. For additional assurance, some developers perform statistical analysis of the likelihood of code sequences appearing in legitimate programs. Still, it is impossible to rule out false positives. Signatures are simply program segments; therefore, the code could appear in an uninfected program.

False negatives occur when an infected executable is encountered but no pattern match is detected. This usually results from procedural problems; if a stealth virus is memory-resident at the time the scanner executes, the virus may hide itself. False negatives can also occur when the system has been infected by a virus that was unknown at the time the scanner was built.

Scanners are also prone to misidentification or may lack precision in naming. Misidentification will usually occur when a new variant of an older virus is encountered. As an example, a scanner may proclaim that *Jerusalem-B* has been detected, when in fact the *Jerusalem-Groen Links* virus is present. This can occur because these viruses are both *Jerusalem* variants and share much of their code. Another scanner might simply declare "Jerusalem variant found in *filename*." This is accurate, but rather imprecise.

Ease of Use

Scanners are very easy to use in general. You simply execute the scanner and it provides concise results. The scanner may have a few options describing which disk, files, or directories to scan, but the user does not have to be a computer expert to select the right parameters or comprehend the results.

Administrative Overhead

New viruses are discovered every week. As a result, virus scanners are immediately out of date. If an organization distributes scanners to its users for virus detection, procedures must be devised for distribution of updates. A scanner for a DOS PC that is more than a few months old will not detect most newly developed viruses. (It may detect, but misidentify, some new variants.) Timely updates are crucial to the effectiveness of any scanner-based anti-virus solution. This can present a distribution problem for a large organization.

Installation is generally simple enough for any user to perform. Interpreting the results is very simple when viruses are correctly identified. Handling false positives will usually require some assistance from technical support. This level of support may be available from the vendor.

Efficiency

Scanners are very efficient. There is a large body of knowledge about searching algorithms, so the typical scanner executes very rapidly. Proactive application will generally result in higher system overhead.

4.1.3 Summary

Scanners are extremely effective at detecting known viruses. Scanners are not intended to detect new viruses (i.e., any virus discovered after the program was released) and any such detection will result in misidentification. Scanners enjoy an especially high level of user acceptance because they name the virus or virus family. However, this can be undermined by the occurrence of false positives.

The strength of a scanner is highly dependent upon the quality and timeliness of the signature database. For viruses requiring algorithmic methods, the quality of the algorithms used will be crucial.

The major strengths of scanners are:

- Up-to-date scanners can be used to reliably detect more than 95 percent of all virus infections at any given time.

- Scanners identify both the infected executable and the virus that has infected it. This can speed the recovery process.

- Scanners are an established technology, utilizing highly efficient algorithms.

- Effective use of scanners usually does not require any special knowledge of the computer system.

The major limitations of scanners are:

- A scanners only looks for viruses that were known at the time its database of signatures was developed. As a result, scanners are prone to false negatives. The user interprets "No virus detected" as "No virus exists." These are *not* equivalent statements.

- Scanners *must* be updated regularly to remain effective. Distribution of updates can be a difficult and time-consuming process.

- Scanners do not perform precise identification. As a result, they are prone to false positives and misidentification.

4.2 General Purpose Monitors

General purpose monitors protect a system from the replication of viruses or execution of the payload of Trojan horses by actively intercepting malicious actions.

4.2.1   Functionality

Monitoring programs are active tools for the real-time detection of viruses and Trojan horses. These tools are intended to intervene or sound an alarm every time a software package performs some suspicious action considered to be virus-like or otherwise malicious behavior. However, since a virus is a code stream, there is a very real possibility that legitimate programs will perform the same actions, causing the alarms to sound.

The designer of such a system begins with a model of "malicious" behavior, then builds modules which intercept and halt attempts to perform those actions. Those modules operate as a part of the operating system.

4.2.2   Selection Factors

Accuracy

A monitoring program assumes that viruses perform actions that are in its model of suspicious behavior and in a way that it can detect. These are not always valid assumptions. New viruses may utilize new methods which may fall outside of the model. Such a virus would not be detected by the monitoring program.

The techniques used by monitoring tools to detect virus-like behavior are also not foolproof. Personal computers lack memory protection, so a program can usually circumvent any control feature of the operating system. As a part of the operating system, monitoring programs are vulnerable to this as well. There are some viruses which evade or turn off monitoring programs.

Finally, legitimate programs may perform actions that the monitor deems suspicious (e.g., self-modifying programs).

Ease of Use

Monitoring software is not appropriate for the average user. The monitor may be difficult to configure properly. The rate of false alarms can be high, particularly false positives, if the configuration is not optimal.

The average user may not be able to determine that program A should modify files, but program B should not. The high rate of false alarms can discourage such a user. At worst, the monitor will be turned off or ignored altogether.

<u>Administrative Overhead</u>

Monitoring programs can impose a fairly heavy administrative workload. They impose a moderate degree of overhead at installation time; this is especially true if several different systems are to be protected. The greatest amount of overhead will probably result from false positives, though. This will vary greatly according to the users' level of expertise.

On the other hand, the monitoring software does not have to be updated frequently. It is not virus-specific, so it will not require updating until new virus *techniques* are devised. (It is still important to remain up-to-date; each time a new class of virus technology is developed, a number of variations emerge.)

<u>Efficiency</u>

Monitoring packages are integrated with the operating system so that additional security procedures are performed. This implies some amount of overhead when any program is executed. The overhead is usually minimal, though.

4.2.3   Summary

Monitoring software may be difficult to use but may detect some new viruses that scanning does not detect, especially if they do not use new techniques.

These monitors produce a high rate of false positives. The users of these programs should be equipped to sort out these false positives on their own. Otherwise, the support staff will be severely taxed.

Monitors can also produce false negatives if the virus doesn't perform any activities the monitor deems suspicious. Worse yet, some viruses have succeeded in attacking monitored systems by turning off the monitors themselves.

4.3   Access Control Shells

Access control shells function as part of the operating system, much like monitoring tools. Rather than monitoring for virus-like behavior, the shell attempts to enforce an access control policy for the system. This policy is described in terms of programs and the data files they may access. The access control shell will sound an alarm every time a user attempts to access or modify a file with an unauthorized software package.

4.3.1   Functionality

To perform this process, the shell must have access to identification and authentication information. If the system does not provide that information, the access control shell may include it. The access control shell may also include encryption tools. These tools can be used to ensure that a user does not reboot from another version of the operating system to circumvent the controls. Note that may of these tools require additional hardware to accomplish these functions.

Access control shells are policy enforcement tools. As a side benefit, they can perform real-time detection of viruses and Trojan horses. The administrator of such a system begins with a description of authorized system use, then converts that description into a set of critical files and the programs which may be used to modify them. The administrator must also select the files which require encryption.

For instance, a shipping clerk might be authorized to access the inventory database with a particular program. However, that same clerk may not be allowed to access the database directly with the database management software. The clerk may not be authorized to access the audit records generated by the trusted application with any program. The administrator would supply appropriate access control statements as input to the monitor and might also encrypt the database.

4.3.2   Selection Factors

<u>Accuracy</u>

Access control shells, like monitoring tools, depend upon the virus or Trojan horse working in an expected manner. On personal computer systems, this is not always a valid assumption. If the virus uses methods that the access control shell does not monitor, the monitor will produce false negatives.

Even with the access control shell, a well-behaved virus can modify any program that its host program is authorized to modify. To reduce the overhead, many programs will not be specifically constrained. This will allow a virus to replicate and is another source of false negatives.

False positives can also occur with access control shells. The system administrator must have sufficient familiarity with the software to authorize access to every file the software needs. If not, legitimate accesses will cause false alarms. If the system is stable, such false positives should not occur after an initial debugging period.

Ease of Use

These tools are intended for highly constrained environments. They usually are not appropriate for the average user at home. They can also place a great deal of overhead on system administrators. The access control tables must be rebuilt each time software or hardware is added to a system, job descriptions are altered, or security policies are modified. If the organization tends to be dynamic, such a tool will be very difficult to maintain. Organizations with well-defined security policies and consistent operations may find maintenance quite tolerable.

This software is easy for users, though. They simply log in and execute whatever programs they require against the required data. If the access control shell prevents the operation, they must go through the administrator to obtain additional privileges.

Efficiency

An access control shell modifies the operating system so that additional security procedures are performed. This implies some amount of overhead when any program is executed. That overhead may be substantial if large amounts of data must be decrypted and re-encrypted upon each access.

Administrative Overhead

An access control shell should not require frequent updates. The software is not specific to any particular threat, so the system will not require updates until new techniques are devised for malicious code. On the other hand, the access control tables which drive the software may require frequent updates.

4.3.3 Summary

Access control shells may be difficult to administer, but are relatively easy for the end-user. This type of tool is primarily designed for policy enforcement, but can also detect the replication of a virus or activation of a Trojan horse.

The tool may incur high overhead processing costs or be expensive due to hardware components. Both false positives and false negatives may occur. False positives will occur when the access tables do not accurately reflect system processing requirements. False negatives will occur when virus replication does not conflict with the user's access table entries.

4.4   Checksums for Change Detection

Change detection is a powerful technique for the detection of viruses and Trojan horses. Change detection works on the theory that executables are static objects; therefore, modification of an executable implies a possible virus infection. The theory has a basic flaw: some executables are self-modifying. Additionally, in a software development environment, executables may be modified by recompilation. These are two examples where checksumming may be an inappropriate solution to the virus problem.

4.4.1   Functionality

Change detection programs generally use an executable as the input to a mathematical function, producing a *checksum*. The change detection program is executed once on the (theoretically) clean system to provide a baseline[3] for testing. During subsequent executions, the program compares the computed checksum with the baseline checksum. A change in the checksum indicates a modification of the executable.

Change detection tools are reactive virus detection tools. They can be used to detect any virus, since they look for modifications in executables. This is a requirement for any virus to replicate. As long as the change detector reviews every executable in its entirety on the system and is used in a proper manner, a virus cannot escape detection.

Change detection tools employ two basic mathematical techniques: Cyclic Redundancy Checks (CRC) and cryptographic checksums.

CRC-Codings

CRC checksums are commonly used to verify integrity of packets in networks and other types of communications between computers. They are fairly efficient and well understood. CRC-based checksums are not extremely secure; they are based on a known set of algorithms. Therefore they can be broken (the particular algorithm can be guessed) by a program if it can find the checksum for a file.

CRC checksum tools, like all change detection tools, can only detect that a virus has replicated. Additionally, the executable must be appear in the baseline.

Cryptographic Checksums

Cryptographic checksums are obtained by applying cryptographic algorithms to the data. Both public and private key algorithms can be used. In general, private key algorithms are

---

[3]The original file names and their corresponding checksums.

used for efficiency. These techniques are sometimes used in conjunction with two other procedures to decrease system overhead. These techniques are message digesting and hashing.[4]

In *Message Digesting*, hashing is used in conjunction with cryptographic checksums. The hash function, which is very fast, is applied directly to the executable. The result is much smaller than the original data. The checksum is computed by applying the cryptographic function to the hash result. The final result approaches the cryptographic checksum for security, but is much more efficient.

4.4.2   Selection Factors

Accuracy

Properly implemented and used, change detection programs should detect every virus. That is, there are no false negatives with change detection. Change detection can result in high numbers of false positives, however. Programs tend to store configuration information in files containing executable code. If these files are checksummed, as they should be, a change in configuration will trigger the change detector. Additionally, the system must be virus-free when the checksums are calculated; resident viruses may fool the change detection software.

Ease of Use

Change detection software is more challenging to use than some other anti-virus tools. It requires good security procedures and substantial knowledge of the computer system. Procedurally, it is important to protect the baseline. The checksums should be stored off-line or encrypted. Manipulation of the baseline will make the system appear to have been attacked.

Analysis of the results of a checksumming procedure is also more difficult. The average user may not be able to determine that one executable is self-modifying but another is not. False positives due to self-modifying code can discourage such a user, until the output of the change detector is ignored altogether.

Administrative Overhead

Change detection software is easy to install and it requires no updates. The baseline must be established by a qualified staff member. This includes the initial baseline, as well as changes to the baseline as programs are added to the system. Once in operation, a high degree of support can be required for the average end-user, however. A qualified staff member must be available to determine whether or not a change to a particular executable is due to a virus or simply a result of self-modification.

---

[4]Discussion of cryptographic terminology is beyond the scope of this document. Please see [Sim92]

Efficiency

Change detectors do not impose any overhead on general system use. There is, however, some storage overhead for the baseline checksums. These are best stored off-line with the checksum program.

The calculation of checksums is computationally intensive; the mathematical functions must be calculated on at least a portion of the executable. To be exhaustive, the function should be calculated on the entire executable.

4.4.3   Summary

If change is detected, there are several possibilities: a virus infection, self-modification, recompilation, or modification of the baseline. A knowledgeable user is required to determine the specific reason for change.

The primary strength of change detection techniques is the ability to detect new viruses and Trojan horses. The limitation of change detection is the need for a knowledgeable user to interpret the output.

4.5   Knowledge-Based Virus Removal Tools

The primary means of automated removal of virus infection is knowledge-based removal tools. These removal tools attempt to reverse the modifications a virus makes to a file. After analyzing a particular virus to determine its effects on an infected file, a suitable algorithm is developed for disinfecting files. Tools are available which address only a single virus. These single virus disinfectors are usually developed as the result of a particularly virulent outbreak of a virus. Others detectors are general virus removal programs, containing removal algorithms for several viruses.

4.5.1   Functionality

Knowledge-based removal tools restore an executable to its pre-infection state. All modifications to the original executable must be known in order to accomplish this task. For example, if a file is infected with an overwritting virus, removal is not possible. The information that was overwritten cannot be restored.

The most critical piece of information in the removal process is the identity of the virus itself. If the removal program is removing *Jerusalem-DC*, but the host is infected with *Jerusalem-E2*, the process could fail. Unfortunately, this information is often unavailable or imprecise. This is why precise identification tools are needed.

4.5.2   Selection Factors

Disinfecting software is not very accurate, for a variety of reasons. The error rates are fairly high; however, most are soft errors. This is a result of incomplete information regarding the virus and the lack of quality assurance among virus writers. Additionally, removal techniques tend to fail when a system or file has been infected multiple times (i.e., by the same virus more than once, or by more than one virus).

These programs are relatively easy to use and can disinfect large numbers of programs in a very short time. Any system overhead is inconsequential since the system should not be used until the virus is removed.

4.5.3   Summary

Accurate removal may not be possible. Even if it is theoretically possible, precise identification of the virus is necessary to ensure that the correct removal algorithm is used.

Certain viruses (e.g., overwriting viruses) always cause irreparable damage to an executable. Some extraordinarily well-behaved viruses can be disinfected every time. Most viruses fall somewhere in between. Disinfection will often work, but the results are unpredictable.

Some executables cannot be recovered to the exact pre-infection state. In such a case, the file length or checksum of the disinfected executable may differ from the pre-infection state. In such a case, it is impossible to predict the behavior of the disinfected program. This is the reason virus researchers generally dislike removal programs and discourage their use.

4.6   Research Efforts

The following subsections describe research areas in the anti-virus field. New tools, based on techniques developed in these and other areas, may be available in the near future.

4.6.1   Heuristic Binary Analysis

Static analysis detection tools, based upon heuristic binary analysis, are a focus of research at this time. Heuristic binary analysis is a method whereby the analyzer traces through an executable looking for suspicious, virus-like behavior. If the program appears to perform virus-like actions, a warning is displayed.

Functionality

Binary analysis tools examine an executable for virus-like code. If the code utilizes techniques which are common to viruses, but odd for legitimate programs, the executable is flagged as "possibly infected." Examples include self-encrypted code or code that appears to have been appended to an existing program.

Selection Factors

Both false positives and negatives are sure to result with use of this type of software. False positives occur when an uninfected program uses techniques common to viruses but uncommon in legitimate programs. False negatives will occur when virus code avoids use of those techniques common to viruses.

Binary analysis tools are fairly easy to use. The user simply specifies a program or directory to be analyzed. Analyzing the results is more difficult. Sorting out the false positives from real infections may require more knowledge and experience than the average user possesses.

Heuristic analysis is more computationally intensive than other static analysis methods. This method would be inappropriate for daily use on a large number of files. It is more appropriate for one-time use on a small number of files, as in acceptance testing.

A heuristic analysis program will require updates as new techniques are implemented by virus writers.

Summary

Early examples of this class of tool appear to have fairly high error rates as compared with commercial detection software. As with system monitors, it is difficult to define suspicious in a way that prevents false positives and false negatives. However, these types of tools have been used successfully to identify executables infected by "new" viruses in a few actual outbreaks.

Heuristic binary analysis is still experimental in nature. Initial results have been sufficiently encouraging to suggest that software acceptance procedures could include these tools to augment more traditional technology.

4.6.2   Precise Identification Tools

Precise identification tools are a means by which viruses are named with a much higher degree of assurance. These tools are intended to augment detection tools. Once a virus has been detected, a precise identification tool would be invoked in order to more accurately identify the virus.

Functionality

Virus scanners, currently the most common virus detection method, generally employ signature scanning to detect and identify viruses. This method, however, can lead to misidentifications. The signature that the scanner matched could appear in more than one variant of the virus. To avoid mis-identification the whole virus must match, not just a subset of the virus (i.e., the signature). It is neither feasible nor desirable for identification software to be distributed containing the code to all viruses it can detect. Therefore, prototype precise identification tools utilize a "virus map" to represent the contents of the virus. The virus map contains checksum values for all constant parts of the virus code. The map skips over sections of the virus that contain variable information such as text or system dependent data values.

If the checksums generated by the corresponding portions of the program match, the program is almost certainly infected by the virus corresponding to the map. If none of the maps in the database correspond, the program is infected by a new virus (or is uninfected.)

Selection Factors

The quality of the results produced by a precise identification tool is dependent upon the quality of the virus map database. If that has been done well and kept current, these tools are extremely accurate and precise when identifying known viruses. Conversely, if the virus is new or has no corresponding entry in the database, the precise identification tool should always "fail" to identify the viruses.

This type of tool is easy to use. The user simply specifies an executable, and the tool returns a name, if known. The results are straightforward; it is virus "X," or unknown.

Precise identification tools are slow due to the intensive nature of the computations. These tools may be used to perform an identification pass after the use of a more efficient detection tool. Such a plan would provide the user with the benefits of precise identification without great overhead. Once a virus has been detected, the user wants to know exactly what virus he has and time is not a significant factor.

Summary

Users want to know more about the virus infecting their systems. Precise identification will help them obtain more complete information and can also facilitate automated removal.

Researchers will also wish to use this type of tool. It will allow them to separate samples of known viruses from new ones without performing analysis.

4.7   Other Tools

The remaining tools, system utilities and inoculation, are included for completeness. These tools can be used to provide some measure of functionality. In general, however, these tools are weaker than general anti-virus tools.

4.7.1   System Utilities

Some viruses can be detected or removed with basic system utilities.[5] For example, most DOS boot sector infectors and some Macintosh viruses can be removed with system utilities. System utilities can also be used to detect viruses by searching for virus signatures. These tools have a rather limited focus, though.

Viruses that can be disinfected "by hand" are generally the extremely well-behaved, highly predictable viruses that are well understood. Such viruses are the exception, not the rule. There are many more viruses that cannot be disinfected with these tools.

Where possible, disinfection with system utilities will produce dependable results. A reasonable amount of knowledge is required about the computer system and the virus itself, though. This technique can also be very laborious if a large number of systems are infected.

System utilities are an inefficient means of detection. Generally, only one signature can be handled at a time. This might be a useful technique if a specific virus is to be detected.

Summary

Accurate removal by system utilities is frequently impossible. Certain classes of viruses (e.g., overwriting viruses) always damage the executable beyond all hope of repair. Others modify the executable in rather complicated ways. Only viruses that are extremely well-behaved can be disinfected every time. Similarly, detection with system utilities has limited application.

---

[5]Two examples of these system utilities are Norton Utilities for the PC and ResEdit for the Macintosh.

4.7.2 Inoculation

In some cases, an executable can be protected against a small number of viruses by "inoculation." This technique involves attaching the self-recognition code for the virus to the executable at the appropriate location.

Since viruses may place their self-recognition codes in overlapping locations, the number of viruses that can be inoculated against simultaneously will be small. To make matters worse, a common way to create a new variant is to change the self-recognition code. Thus, this technique will often fail when tested by minor variants of the viruses inoculated against.

Inoculation is no substitute for more robust anti-virus tools and procedures. It *might* be useful, though, if an organization has had recurring infections from a single virus. For example, after cleaning three or four outbreaks of a particular virus from a network of PCs, inoculation might be considered as a desperation measure.

5. Selecting Anti–Virus Techniques

The selection of the appropriate class of anti-virus tools requires answers to the following set of questions:

- What is the probability of a virus infection?
- What are the consequences of a virus infection?
- What is the skill level of the users in your organization?
- What level of support is available to the end-user?

The first two questions address risk; security should always be commensurate with need. The third and fourth questions address the limitations of the tools and personnel. The answers will be different for each person or organization.

Every organization is at some risk of virus infection. Virus infections can occur whenever electronic information is shared. Every organization shares information in some way and is a potential victim of a virus infection. Most organizations should have some tools available to detect such an infection.

Personal computer users may benefit from tools to identify viruses, since so many viruses exist. Identification tools are not necessary where viruses are few or only theoretically possible.

The use of removal tools is generally not required.[6] It may be desirable in situations where a single person or a small team is tasked with cleaning up after an infection or where high connectivity can result in rapid spread of the virus (such as networks).

5.1   Selecting Detection Tools

The first point to consider when selecting a detection product is the type of viruses likely to be encountered. Approximately 95 percent of all virus infections are accounted for by a small number of viruses. The viruses that constitute this small set can vary geographically. The common viruses can be distinct on different continents, due to the paths in which they travel. Of course, different hardware platforms will be at risk from different viruses.

International organizations may be vulnerable to a larger set of viruses. This set may be obtained by merging the sets of viruses from different geographical regions where they do

---

[6]Exceptions, such as the DIR-2 PC virus, may be extremely difficult to remove without appropriate tools. In this case, the only alternative to removal tools is to format the disk.

business. Organizations with contacts or installations in locations where virus writers are particularly active [Bon91] are also more likely to encounter new viruses.

Risk from new viruses is an important consideration. Scanners are limited by their design to known viruses; other detection tools are designed to detect any virus. If your organization is at high risk from new viruses, scanners should not be the sole detection technique employed.

Another important criteria to consider is the number and type of errors considered tolerable. The tolerance for a particular type of error in an organization will vary according to the application. Table 1 shows the types of errors which should be expected. An estimate of the frequency that this class of error is encountered (*Infrequent*, *Frequent*, or *Never*) is also given for each class of tools and error type. All anti-virus tools are subject to errors, but their relative frequencies vary widely. Scanners probably have the lowest overall error rate. Checksummers do not produce false negatives.

Detection Tool / Error Types	Scanner	Checksum	Binary Analysis	Generic Monitor	Access Control Shell
False Positives	Infrequent Signatures can occur in valid files	Frequent Every time a program is modified	Frequent In our test, 15% errors	Frequent Whenever a legitimate program performs virus-like actions	Frequent Whenever a legitimate program performs virus-like actions
False Negatives	Infrequent May not detect variants; won't detect new viruses	Never Viruses always change executables	Frequent In our test, 8% errors	Frequent Viruses that circumvent OS can be missed	Frequent Viruses that circumvent OS can be missed

Table 1: Types of errors.

The third and fourth items to consider when selecting anti-virus tools are the ease of use and administrative overhead required for each tool. Questions to consider are:

- What is the average skill level of your organization's end-user?
- Does your organization have a support staff to assist user with more technical problems?

Table 2 includes a general evaluation of the ease of use and administrative overhead imposed by each class of tools.

Detection Tool / Criteria	Scanner	Checksum	Binary Analysis	Generic Monitor	Access Control Shell
Easy of Use	**Very Good** Requires no special knowledge of the system	**Average** Easy to use; results may be difficult to interpret	**Poor** Easy to use; results must be verified	**Poor** Results are difficult to interpret	**Average** Easy to use; Configuration is an impediment
Administrative Overhead	**Low** Requires frequent updates. Little add'l support req'd	**Low** No updates req. Assist in interpreting results	**High** Few updates. Much verification of results	**High** Few updates. Much verification of results	**High** Few updates. Much verification of results

Table 2: Personnel requirements.

If several tools still appear to be candidates, consider the functionality of these tools *beyond* virus detection. Viruses are only one of the many threats to computer security. All detection tools except scanners have general security applications beyond viruses. Scanners are limited in application to viruses, but have the added functionality of virus identification.[7] Consider the added functionality which is most needed by your organization and choose accordingly. The alternatives are outlined in table 3.

	Detection Tool				
	Scanner	Checksum	Binary Analysis	Generic Monitor	Access Control Shell
Additional Functionality	Identification; May also detect known trojan horses	Detection of trojan horses and altered data	Detection of trojan horses	Detection of trojan horses	Enforcing organizations security policy

Table 3: Additional functionality.

The final selection criteria to be considered is when does the tool detect viruses. Proactive detection tools allow the user to keep viruses off a system by testing incoming software. These

---

[7]Some scanners can also detect known Trojan horses.

tools only allow one chance of detecting a virus (upon initial introduction to the system). Active detection tools intervene during the replication phase itself. Reactive detection tools can be used any time after a virus has entered the system. Additionally, reactive tools are not as rigorous in their demands on system performance. Table 4 shows when these different tools detect viruses.

Detection Tool / Point of Detection	Scanner	Checksum	Binary Analysis	Generic Monitor	Access Control Shell
Static Executable	Yes	No	Yes	No	No
Replication Phase	No	No	No	Yes	Yes
After Infection	Yes	Yes	Yes	No	Yes

Table 4: When tools detect?

5.1.1 Combining Detection Tools

The most complete protection will be obtained by combining tools which perform in radically different fashion and protect against different classes of viruses. For instance, when used together a scanner and a checksum program will protect against both known and unknown viruses. The scanner can detect known viruses before software is installed on the system. A virus can be modified to elude the scanner, but it will be detected by the checksum program.

The two tools should have different "additional functionality" (see table 3) to form the most comprehensive security package. For instance, the combination of a checksum program and an access control shell would also detect Trojan horses and enforce organizational security policy in addition to virus detection. On the other hand, adding a binary analyzer to a system that already employs checksumming would not provide additional functionality.

If you must use two scanners, be sure that they use different search strings. A number of tools are based on published search strings; shareware tools commonly utilize the same public domain signature databases. Two different scanner engines looking for the same strings do not provide any additional protection of information.[8]

---

[8]Algorithms for detection tend to be independently developed.

5.2 Identification Tools

Currently, scanners are the only effective means of identifying viruses. As discussed in Section 3.1.2, the accuracy to which scanners identify viruses can vary. In the future, precise identification tools should offer greatly increased accuracy.

5.3 Removal Tools

The most dependable technique for virus removal continues to be deletion of the infected executable and restoration from a clean backup. If backups are performed regularly and in a proper manner, virus removal tools may be neglected.

In large organizations with high connectivity, automated removal tools should be obtained. Virus eradication through the removal of infected executables may require too much time and effort. Knowledge based tools will disinfect the largest number of different viruses, but proper identification of the virus prior to disinfection is critical. Even with knowledge based removal tools, disinfection of executables is not always reliable (see Sec. 3.1.3). Test all disinfected executables to be sure they appear to execute properly. There is still a chance, however, that soft errors will occur.

5.4 Example Applications of Anti-Virus Tools

This section provides hypothetical scenarios for the use of anti-virus tools. For each application, a battery of tools is suggested. There are several ways these tools can be applied to the same scenario; this text represents just one set of rational solutions.

5.4.1 Average End-User

Detailed knowledge of the computer system is not required for the average end-user to perform one's job. Such a user should not be required to obtain detailed knowledge just to use anti-virus tools. This implies that scanners are probably most appropriate for the average end-users. Any other choice will require support from a technical support team or computer security incident response team. Of the remaining tools, the best option is a checksum program. By executing the checksum program regularly, for example weekly or monthly, infections will be detected within a limited timeframe.

Another possibility is to relieve these users of the responsibility of detecting viruses entirely. If a technical support team is already providing other regular services (e.g., backup), the support team can use any combination of anti-virus tools deemed necessary.

5.4.2   Power Users

Power users, those with detailed knowledge of their computer systems, will be better equipped to handle a larger variety of anti-virus tools. A power user is more able to determine whether a change detected by a checksum program is in fact legitimate. Additionally, a power user is going to be better equipped to configure some of the other tools, such as general purpose monitors and access control shells.

5.4.3   Constrained User

If the user is constrained by policy to run a small set of programs against a known set of data files, an access control shell may be the appropriate choice. As an example, consider a data entry clerk who is permitted to run one particular database application and a basic set of utilities: mail, word processing, and a calendar program. An access control shell can be configured so that any changes to executable files by that user are deemed illegal operations. Additionally, if the set of executable files is restricted for the user, it is difficult to introduce a virus into the system. The virus is unable to spread if it can never be executed.

5.4.4   Acceptance Testing

Acceptance testing is a means by which software is verified to be "virus-free" before it is put into daily use. This is usually accomplished by placing the software on an isolated system and performing tests that are intended to mimic every day use. A combination of anti-virus tools is required to adequately perform this function, which must detect both known and future viruses. In particular, a checksum program is most useful. Even if the trigger conditions for the payload are not met, the virus will still most likely attempt to replicate. It is the result of the replication process that a checksum program detects.

5.4.5   Multi-User Systems

Although viruses found in the wild have been limited to personal computer systems, viruses for multi-user systems have been demonstrated in a number of laboratory experiments. Therefore, the potential exists for viruses on multi-user systems. As a result, it is prudent to ensure that the security measures taken on a multi-user system address viruses as well.

Currently, administrators of multi-user systems have a limited number of options for virus protection. Administrators of these systems cannot use monitors or scanners. Since there are no known viruses, there are no signatures to search for or expected virus behavior to detect. An option that is available to administrators of multi-user systems is change detection. Many of these systems are already equipped with a checksum program. Access control shells are another possibility for many systems. Like access control, though, they are not usually designed for virus detection.

5.4.6   Network Server

Network servers present an interesting problem. They can support a wide variety of machines, but may run an entirely different operating system. For instance, a UNIX server may support a network of PC and Macintosh workstations.

The UNIX system cannot be infected by the Jerusalem-B or WDEF viruses, but infected files may be stored on its disk. Once the network server has infected files on it, the workstations it supports will rapidly become infected as well.

Since the viruses never execute on the server, the administrator is limited to static detection techniques such as scanners or change detectors. The nature of network servers allows these tools to be run automatically during off-peak periods.

6. Selecting the Right Tool

Once an anti-virus technique has been selected, an appropriate tool from that class must be selected. This section presents several features to be considered when selecting a specific product from a class of tools.

6.1   Selecting a Scanner

Scanners are implemented in several forms. Hardware implementations, available as add-on boards, scan all bus transfers. Software implementations include both non-resident and resident software for the automatic scanning of diskettes.

Non-resident software is sufficiently flexible to meet most needs; however, to be effective the user must execute the software regularly. Hardware or resident software are better choices for enforcing security policy compliance. Resident scanners may be susceptible to stealth viruses.

Although most scanners use similar detection techniques, notable differences among products exist. Questions that potential users should consider when selecting a scanner include:

- How frequently is the tool updated? A scanner must be updated regularly to remain effective. How frequently updates are needed depends on which platform the scanner is used. Update frequency should be proportional to the rate at which new viruses are discovered on that platform.

- Can the user add new signatures? This can be very important if a particularly harmful virus emerges between updates.

- Does the tool employ algorithmic detection? For which viruses does the tool use algorithmic detection? Algorithmic detection is preferable to the use of multiple signatures to detect polymorphic viruses.

- How efficient is the tool? Users are less likely to use a slow scanner. There can be a significant difference in performance between different search algorithms.

- Does the vendor develop their own virus signatures, or are the signatures based on published search strings? There is nothing particularly wrong with published search strings, but it indicates the level of resources the vendor has committed to the product.

- What is the level of documentation? Some packages arrive with large fact-filled binders; other packages are a single floppy disk with a few ASCII files describing installation and parameters.

6.2   Selecting a General Purpose Monitor

General purpose monitors are usually implemented in software; however, hardware implementations do exist. Hardware versions may be more difficult to circumvent, but they are not foolproof. The following questions should be considered when selecting a general purpose monitor:

- How flexible are the configuration files? Can different parts of the monitor be disabled? Can the monitor be configured so that certain executables can perform suspect actions? For example, a self-modifying executable will still need to be able to modify itself.
- What types of suspect behavior are monitored? The more types of behavior monitored, the better. A flexible configuration to select from the set of features is desirable.
- Can the monitor be reconfigured to scan for additional virus techniques? Are updates provided as new virus techniques are discovered?

6.3   Selecting an Access Control Shell

Access control shells may be implemented in software or as hybrid packages with both hardware and software components. If encryption modules are required, they can be designed as software or hardware. The following questions should be considered when selecting an access control shell:

- What type of access control mechanism does the shell provide and does it fit your security policy?
- If encryption is employed, what is the strength of the algorithms used? In general, publicly scrutinized algorithms are to be preferable to secret, proprietary algorithms where you are depending on the secrecy of the algorithm, rather than secrecy of the key.
- How strong are the identification and authentication mechanisms? [FIP85] provides basic criteria for analyzing the strength of these mechanisms.
- Are the passwords themselves adequately protected? Passwords should never be stored in cleartext.

6.4   Selecting a Change Detector

Due to cost considerations, change detection tools are usually implemented in software. However, hardware implementations do speed the calculation of cryptographic checksums. The following questions should be considered when selecting a change detector:

- What kind of checksum algorithm does the tool use - CRC or cryptographic? CRC algorithms are faster. Cryptographic checksums are more secure.
- Can the tool be configured to skip executables that are known to be self-modifying? Consistent false positives will eventually cause the end-user to ignore the reports.
- How are the checksums stored? Some tools create a checksum file for every executable, which tends to clutter the file system and wastes disk space. Other tools store all checksums in a single file. Not only is this technique a more efficient use of disk space, but it also allows the user to store the checksum file off-line (e.g., on a floppy).

6.5   Selecting an Identification Tool

The following questions should be considered when selecting a scanner for identification:

- How many viruses does it detect? How many different viruses are identified? The former asks how many different viruses are detected, whereas the latter asks how many different names are assigned to these different viruses. If a scanner is using signature strings, signatures can appear in variants. These questions will give some understanding regarding the level of precision provided by a particular tool.
- What names are used by the identification tool? Many viruses have numerous "aliases," so different scanners will produce different names for the same infection. This is especially true with IBM PC viruses. The identification feature of the scanner is only useful if the scanner comes with a virus catalog or uses the same nameset as an available catalog.

Precise identification tools will be more useful when they become available, although the same limitations regarding a virus information catalog will still apply.

6.6   Selecting a Removal Tool

Removal tools are more difficult to evaluate, but the following items may be of assistance:

- Ask for a list of viruses that can be removed, and the general level of accuracy. (For example, "75% of disinfections will result in a working executable.") Ask for a list of viruses that cannot be removed. Use the ratio for the basis of a rough comparison.
- Get a scanner and removal tool that work from the same naming space. The removal tool works on the basis of the virus you name. You need to supply it with the name by which it knows the virus. Matched identification and removal tools are required to make it work.

7. For Additional Information

The National Institute of Standards and Technology's Computer Security Division maintains an electronic bulletin board system (BBS) focusing on information systems security issues. It is intended to encourage sharing of information that will help users and managers better protect their data and systems. The BBS contains the following types of information specific to the virus field:

- alerts regarding new viruses, Trojan horses, and other threats;
- anti-virus product reviews (IBM PC and Macintosh);
- technical papers on viruses, worms, and other threats;
- anti-virus freeware and shareware; and
- archives of the VIRUS-L forum.

Occasionally, the alerts contain signature strings to update scanners. The anti-virus product reviews examine and evaluate specific tools. The papers provide an extensive body of basic knowledge regarding these threats. The VIRUS-L forum has served as a world-wide discussion forum for the exchange of information regarding viruses since April 1988. The past issues are available for download.

Access Information

The NIST Computer Security Resource Center BBS can be access via dial-up or through the Internet via telnet:

Dial-up access: (301) 948-5717 (2400 baud or less)
(301) 948-5140 (9600 baud)
Internet: telnet *cs-bbs.ncsl.nist.gov* (129.6.54.30)

References

[Bon91] Vesselin Bontchev. The bulgarian and soviet virus factories. In *Proceedings of the First International Virus Bulletin Conference*, 1991.

[BP92] Lawrence E. Bassham III and W. Timothy Polk. Precise identification of computer viruses. In *Proceedings of the 15th National Computer Security Conference*, 1992.

[Coh92] Dr. Frederick Cohen. Current best practices against computer viruses with examples from the DOS operating system. In *Proceedings of the Fifth International Computer Virus & Security Conference*, 1992.

[FIP85] Password Usage. Federal Information Processing Standard (FIPS PUB) 112, National Institute of Standards and Technology, May 1985.

[Rad91] Yisrael Radai. Checksumming techniques for anti-viral purposes. In *Proceedings of the First International Virus Bulletin Conference*, 1991.

[Sim92] Gustavus J. Simmons, editor. *Contemporary Cryptology: The Science of Information Integrity*. IEEE Press, 1992.

[Sku92] Fridrik Skulason. The mutation engine - the final nail? *Virus Bulletin*, pages 11–12, April 1992.

[Sol92] Dr. Alan Solomon. Mechanisms of stealth. In *Proceedings of the Fifth International Computer Virus & Security Conference*, 1992.

[WC89] John Wack and Lisa Carnahan. Computer Viruses and Related Threats: A Management Guide. Special Publication 500-166, National Institute of Standards and Technology, August 1989.

Part II

Computer Viruses and
Related Threats
A Management Guide

Executive Summary

Computer viruses and related threats represent an increasingly serious security problem in computing systems and networks. This document presents guidelines for preventing, deterring, containing, and recovering from attacks of viruses and related threats. This section acquaints senior management with the nature of the problem and outlines some of the steps that can be taken to reduce an organization's vulnerability.

What Are Computer Viruses and Related Threats?

Computer viruses are the most widely recognized example of a class of programs written to cause some form of intentional damage to computer systems or networks. A computer virus performs two basic functions: it copies itself to other programs, thereby *infecting* them, and it executes the instructions the author has included in it. Depending on the author's motives, a program infected with a virus may cause damage immediately upon its execution, or it may wait until a certain event has occurred, such as a particular date and time. The damage can vary widely, and can be so extensive as to require the complete rebuilding of all system software and data. Because viruses can spread rapidly to other programs and systems, the damage can multiply geometrically.

Related threats include other forms of destructive programs such as Trojan horses and network worms. Collectively, they are sometimes referred to as *malicious software*. These programs are often written to masquerade as useful programs, so that users are induced into copying them and sharing them with friends and work colleagues. The malicious software phenomena is fundamentally a people problem, as it is authored and initially spread by individuals who use systems in an unauthorized manner. Thus, the threat of unauthorized use, by unauthorized *and* authorized users, must be addressed as a part of virus prevention.

What Are the Vulnerabilities They Exploit?

Unauthorized users and malicious software may gain access to systems through inadequate system security mechanisms, through security holes in applications or systems, and through weaknesses in computer management, such as the failure to properly use existing security mechanisms. Malicious software can be copied intentionally onto systems, or be spread when users unwittingly copy and share infected software obtained from public software repositories, such as software bulletin boards and shareware. Because malicious software often hides its destructive nature by performing or claiming to perform some useful function, users generally don't suspect that they are copying and spreading the problem.

Why Are Incidents of Viruses and Related Threats On the Rise?

Viruses and related threats, while not a recent phenomena, have had relatively little attention focused on them in the past. They occurred less frequently and caused relatively little damage. For these reasons, they were frequently treated lightly in computer design and by management, even though their potential for harm was known to be great.

Computer users have become increasingly proficient and sophisticated. Software applications are increasingly complex, making their bugs and security loopholes more difficult to initially detect and correct by the manufacturer. In conjunction with these two factors, some brands of software are now widely used, thus their bugs and security loopholes are often known to users. With the widespread use of personal computers that lack effective security mechanisms, it is relatively easy for knowledgeable users to author malicious software and then dupe unsuspecting users into copying it.

Steps Toward Reducing Risk

Organizations can take steps to reduce their risk to viruses and related threats. Some of the more important steps are outlined below.

- Include the damage potential of viruses, unauthorized use, and related threats in risk analysis and contingency planning. Develop a plan to deal with potential incidents.

- Make computer security education a prerequisite to any computer use. Teach users how to protect their systems and detect evidence of tampering or unusual activity.

- Ensure that technically oriented security and management staff are in place to deal with security incidents.

- Use the security mechanisms that exist in your current software. Ensure that they are used correctly. Add to them as necessary.

- Purchase and use software tools to aid in auditing computing activity and detecting the presence of tampering and damage.

1. Introduction

This document provides guidance for technical managers for the reduction of risk to their computer systems and networks from attack by computer viruses, unauthorized users, and related threats. The guidance discusses the combined use of policies, procedures, and controls to address security vulnerabilities that can leave systems open to attack. The aim of this document is not to provide solutions to the wide range of specific problems or vulnerabilities, rather it is to help technical managers administer their systems and networks such that manifestations of viruses and related threats can be initially prevented, detected, and contained.

1.1 Audience and Scope

This document is intended primarily for the managers of multi-user systems, personal computers, and associated networks, and managers of end-user groups. Additionally, the document is useful for the users of such systems. The document presents an overview of computer viruses and related threats, how they typically work, the methods by which they can attack, and the harm they can potentially cause. It then presents guidance in the following areas:

- *Multi-User Systems and Associated Networks* - with guidance directed at managers of medium to small systems (as opposed to mainframes that already provide generally effective security controls or are by their nature more secure) and associated wide area and large local area networks, as well as managers of end-users of such systems

- *Personal Computer Systems and Networks* - guidance is directed at those responsible for the management of personal computers and personal computer networks, as well as the managers of personal computer end-users

Within these general categories, individual computing environments will vary widely, from size of computer to user population to type of software and computing requirements. To accommodate these differences, the guidance presented here is general in nature. It attempts to address computer security problems and vulnerabilities that are likely to be found in most computing environments. This document does not address problems directly related to specific brands of software or hardware. A reading list at the end of the document contains references and pointers to other literature that address specific systems and software.

Recommended control measures are grouped according to categories that include general policies and procedures, education, software management, technical controls, monitoring, and contingency planning.   The guidance emphasizes the need for a strong security program as a means for protection from manifestations of viruses and related threats, and as a means for providing detection, containment, and recovery. Such a security program requires personal involvement on the part of management to ensure that the proper policies, procedures, and technical controls exist, and that users are educated so that they can follow safe computing practices and understand the proper actions to take if they detect the presence of viruses or related threats.   The guidelines recommend that network managers, multi-user system managers, end-users, and end-user managers work with each other and approach virus protection from an organizationally consistent basis.

1.2   How to Use This Guide

This document is divided into five chapters and two appendices.   Chapter 2 describes in general how viruses and related software operate, the vulnerabilities they exploit, and how they can be introduced into systems and networks.   Chapter 3 discusses general protection strategies and control measures that apply to technical and end-user management in general; this is done so that the same guidance need not be repeated for each of the succeeding chapters that deal with specific environments.   Chapters 4 and 5 present guidance specific to multi-user and personal computer environments, respectively.   The guidance in these chapters is directed at the respective technical managers and managers of associated networks, as well as the managers of end-user groups that use such systems and networks.   It is recommended that all readers, regardless of their management perspective, examine Chapters 3, 4, and 5 to gain a fuller appreciation of the whole environment with regard to threats, vulnerabilities, and controls.

Appendix A contains document references, while Appendix B contains a reading list with references to general and specific information on various types of viruses, systems, and protective measures. Readers can use these documents to obtain information specific to their individual systems and software.

2. A Brief Overview on Viruses and Related Threats

The term *computer virus* is often used in a general sense to indicate any software that can cause harm to systems or networks. However, computer viruses are just one example of many different but related forms of software that can act with great speed and power to cause extensive damage - other important examples are Trojan horses and network worms. In this document, the term *malicious software* refers to such software.

2.1 Trojan Horses

A Trojan horse[1] program is a useful or apparently useful program or command procedure containing hidden code that, when invoked, performs some unwanted function. An author of a Trojan horse program might first create or gain access to the source code of a useful program that is attractive to other users, and then add code so that the program performs some harmful function in addition to its useful function. A simple example of a Trojan horse program might be a calculator program that performs functions similar to that of a pocket calculator. When a user invokes the program, it appears to be performing calculations and nothing more, however it may also be quietly deleting the user's files, or performing any number of harmful actions. An example of an even simpler Trojan horse program is one that performs only a harmful function, such as a program that does nothing but delete files. However, it may appear to be a useful program by having a name such as CALCULATOR or something similar to promote acceptability.

Trojan horse programs can be used to accomplish functions indirectly that an unauthorized user could not accomplish directly. For example, a user of a multi-user system who wishes to gain access to other users' files could create a Trojan horse program to circumvent the users' file security mechanisms. The Trojan horse program, when run, changes the invoking user's file permissions so that the files are readable by any user. The author could then induce users to run this program by placing it in a common directory and naming it such that users will think the program is a useful utility. After a user runs the program, the author can then access the information in the user's files, which in this example could be important work or personal information. Affected users may not notice the changes for long periods of time unless they are very observant.

---

[1] named after the use of a hollow wooden horse filled with enemy soldiers used to gain entry into the city of Troy in ancient Greece.

An example of a Trojan horse program that would be very difficult to detect would be a compiler on a multi-user system that has been modified to insert additional code into certain programs as they are compiled, such as a login program.   The code creates a *trap door* in the login program which permits the Trojan horse's author to log onto the system using a special password. Whenever the login program is recompiled, the compiler will always insert the trap door code into the program, thus the Trojan horse code can never be discovered by reading the login program's source code.   For more information on this example, see [THOMPSON84].

Trojan horse programs are introduced into systems in two ways: they are initially planted, and unsuspecting users copy and run them.   They are planted in software repositories that many people can access, such as on personal computer network servers, publicly-accessible directories in a multi-user environment, and software bulletin boards.   Users are then essentially duped into copying Trojan horse programs to their own systems or directories.   If a Trojan horse program performs a useful function and causes no immediate or obvious damage, a user may continue to spread it by sharing the program with other friends and co-workers.   The compiler that copies hidden code to a login program might be an example of a deliberately planted Trojan horse that could be planted by an authorized user of a system, such as a user assigned to maintain compilers and software tools.

2.2 Computer Viruses

Computer viruses, like Trojan horses, are programs that contain hidden code which performs some usually unwanted function.   Whereas the hidden code in a Trojan horse program has been deliberately placed by the program's author, the hidden code in a computer virus program has been added by another program, that program itself being a computer virus or Trojan horse.   Thus, computer viruses are programs that copy their hidden code to other programs, thereby *infecting* them.   Once infected, a program may continue to infect even more programs.   In due time, a computer could be completely overrun as the viruses spread in a geometric manner.

An example illustrating how a computer virus works might be an operating system program for a personal computer, in which an infected version of the operating system exists on a diskette that contains an attractive game.   For the game to operate, the diskette must be used to boot the computer, regardless of whether the computer contains a hard disk with its own copy of the (uninfected) operating system program.   When the computer is booted using the diskette, the infected program is loaded into memory and begins to run.   It immediately searches for other copies of the operating system program, and finds one on the hard disk.   It then copies its hidden code to the program on the hard disk.   This happens so quickly that the user may not notice the slight delay before his game is run. Later, when the computer is booted using the hard disk, the newly infected version of the operating system will be loaded into memory. It will in turn look for copies to infect. However, it may also perform any number of very destructive actions, such as deleting or scrambling all the files on the disk.

A computer virus exhibits three characteristics: a *replication* mechanism, an *activation* mechanism, and an *objective*. The replication mechanism performs the following functions:

- searches for other programs to infect

- when it finds a program, possibly determines whether the program has been previously infected by checking a flag

- inserts the hidden instructions somewhere in the program

- modifies the execution sequence of the program's instructions such that the hidden code will be executed whenever the program is invoked

- possibly creates a flag to indicate that the program has been infected

The flag may be necessary because without it, programs could be repeatedly infected and grow noticeably large. The replication mechanism could also perform other functions to help disguise that the file has been infected, such as resetting the program file's modification date to its previous value, and storing the hidden code within the program so that the program's size remains the same.

The *activation* mechanism checks for the occurrence of some event. When the event occurs, the computer virus executes its *objective*, which is generally some unwanted, harmful action. If the activation mechanism checks for a specific date or time before executing its objective, it is said to contain a *time bomb*. If it checks for a certain action, such as if an infected program has been executed a preset number of times, it is said to contain a *logic bomb*. There may be any number of variations, or there may be no activation mechanism other than the initial execution of the infected program.

As mentioned, the objective is usually some unwanted, possibly destructive event. Previous examples of computer viruses have varied widely in their objectives, with some causing irritating but harmless displays to appear, whereas others have erased or modified files or caused system hardware to behave differently. Generally, the objective consists of whatever actions the author has designed into the virus.

As with Trojan horse programs, computer viruses can be introduced into systems deliberately and by unsuspecting users.   For example, a Trojan horse program whose purpose is to infect other programs could be planted on a software bulletin board that permits users to upload and download programs.   When a user downloads the program and then executes it, the program proceeds to infect other programs in the user's system.   If the computer virus hides itself well, the user may continue to spread it by copying the infected program to other disks, by backing it up, and by sharing it with other users.   Other examples of how computer viruses are introduced include situations where authorized users of systems deliberately plant viruses, often with a time bomb mechanism.   The virus may then activate itself at some later point in time, perhaps when the user is not logged onto the system or perhaps after the user has left the organization.   For more information on computer viruses, see [DENNING88]

2.3 Network Worms

Network worm programs use network connections to spread from system to system, thus network worms attack systems that are linked via communications lines.   Once active within a system, a network worm can behave as a computer virus, or it could implant Trojan horse programs or perform any number of disruptive or destructive actions.   In a sense, network worms are like computer viruses with the ability to infect other systems as well as other programs.   Some people use the term virus to include both cases.

To replicate themselves, network worms use some sort of network vehicle, depending on the type of network and systems.   Examples of network vehicles include (a) a network mail facility, in which a worm can mail a copy of itself to other systems, or (b), a remote execution capability, in which a worm can execute a copy of itself on another system, or (c) a remote login capability, whereby a worm can log into a remote system as a user and then use commands to copy itself from one system to the other.   The new copy of the network worm is then run on the remote system, where it may continue to spread to more systems in a like manner.   Depending on the size of a network, a network worm can spread to many systems in a relatively short amount of time, thus the damage it can cause to one system is multiplied by the number of systems to which it can spread.

A network worm exhibits the same characteristics as a computer virus: a *replication* mechanism, possibly an *activation* mechanism, and an *objective*.   The replication mechanism generally performs the following functions:

- searches for other systems to infect by examining host tables or similar repositories of remote system addresses

- establishes a connection with a remote system, possibly by logging in as a user or using a mail facility or remote execution capability

- copies itself to the remote system and causes the copy to be run

The network worm may also attempt to determine whether a system has previously been infected before copying itself to the system. In a multi-tasking computer, it may also disguise its presence by naming itself as a system process or using some other name that may not be noticed by a system operator.

The activation mechanism might use a time bomb or logic bomb or any number of variations to activate itself. Its objective, like all malicious software, is whatever the author has designed into it. Some network worms have been designed for a useful purpose, such as to perform general "house-cleaning" on networked systems, or to use extra machine cycles on each networked system to perform large amounts of computations not practical on one system. A network worm with a harmful objective could perform a wide range of destructive functions, such as deleting files on each affected computer, or by implanting Trojan horse programs or computer viruses.

Two examples of actual network worms are presented here. The first involved a Trojan horse program that displayed a Christmas tree and a message of good cheer (this happened during the Christmas season). When a user executed this program, it examined network information files which listed the other personal computers that could receive mail from this user. The program then mailed itself to those systems. Users who received this message were invited to run the Christmas tree program themselves, which they did. The network worm thus continued to spread to other systems until the network was nearly saturated with traffic. The network worm did not cause any destructive action other than disrupting communications and causing a loss in productivity [BUNZEL88].

The second example concerns the incident whereby a network worm used the collection of networks known as the Internet to spread itself to several thousands of computers located throughout the United States. This worm spread itself automatically, employing somewhat sophisticated techniques for bypassing the systems' security mechanisms. The worm's replication mechanism accessed the systems by using one of three methods:

- it employed *password cracking*, in which it attempted to log into systems using usernames for passwords, as well as using words from an on-line dictionary

- it exploited a *trap door* mechanism in mail programs which permitted it to send commands to a remote system's command interpreter

- it exploited a *bug* in a network information program which permitted it to access a remote system's command interpreter

By using a combination of these methods, the network worm was able to copy itself to different brands of computers which used similar versions of a widely-used operating system. Many system managers were unable to detect its presence in their systems, thus it spread very quickly, affecting several thousands of computers within two days. Recovery efforts were hampered because many sites disconnected from the network to prevent further infections, thus preventing those sites from receiving network mail that explained how to correct the problems.

It was unclear what the network worm's objective was, as it did not destroy information, steal passwords, or plant viruses or Trojan horses. The potential for destruction was very high, as the worm could have contained code to effect many forms of damage, such as to destroy all files on each system. For more information, see [DENNING89] and [SPAFFORD88].

2.4 Other Related Software Threats

The number of variations of Trojan horses, computer viruses, and network worms is apparently endless. Some have names, such as a *rabbit*, whose objective is to spread wildly within or among other systems and disrupt network traffic, or a *bacterium*, whose objective is to replicate within a system and eat up processor time until computer throughput is halted [DENNING88]. It is likely that many new forms will be created, employing more sophisticated techniques for spreading and causing damage.

2.5 The Threat of Unauthorized Use

In that computer viruses and related forms of malicious software are intriguing issues in themselves, it is important not to overlook that they are created by people, and are fundamentally a people problem. In essence, examples of malicious software are tools that people use to extend and enhance their ability to create mischief and various other forms of damage. Such software can do things that the interactive user often cannot directly effect, such as working with great speed, or maintaining anonymity, or doing things that require programmatic system calls. But in general, malicious software exploits the same vulnerabilities as can knowledgeable users. Thus, any steps taken to reduce the likelihood of attack by malicious software should address the likelihood of unauthorized use by computer users.

3. Virus Prevention in General

To provide general protection from attacks by computer viruses, unauthorized users, and related threats, users and managers need to eliminate or reduce vulnerabilities. A general summary of the vulnerabilities that computer viruses and related threats are most likely to exploit is as follows:

- lack of user awareness - users copy and share infected software, fail to detect signs of virus activity, do not understand proper security techniques

- absence of or inadequate security controls - personal computers generally lack software and hardware security mechanisms that help to prevent and detect unauthorized use, existing controls on multi-user systems can sometimes be surmounted by knowledgeable users

- ineffective use of existing security controls - using easily guessed passwords, failing to use access controls, granting users more access to resources than necessary

- bugs and loopholes in system software - enabling knowledgeable users to break into systems or exceed their authorized privileges

- unauthorized use - unauthorized users can break in to systems, authorized users can exceed levels of privilege and misuse systems

- susceptibility of networks to misuse - networks can provide anonymous access to systems, many are in general only as secure as the systems which use them

As can be seen from this summary, virus prevention requires that many diverse vulnerabilities be addressed. Some of the vulnerabilities can be improved upon significantly, such as security controls that can be added or improved, while others are somewhat inherent in computing, such as the risk that users will not use security controls or follow policies, or the risk of unauthorized use of computers and networks. Thus, it may not be possible to completely protect systems from all virus-like attacks. However, to attain a realistic degree of protection, all areas of vulnerability must be addressed; improving upon some areas at the expense of others will still leave significant holes in security.

To adequately address all areas of vulnerability, the active involvement of individual users, the management structure, and the organization in a *virus prevention program* is essential. Such a program, whether formal or informal, depends on the mutual cooperation of the three groups to identify vulnerabilities, to take steps to correct them, and to monitor the results.

A virus prevention program must be initially based upon effective system computer administration that restricts access to authorized users, ensures that hardware and software are regularly monitored and maintained, makes backups regularly, and maintains contingency procedures for potential problems. Sites that do not maintain a basic computer administration program need to put one into place, regardless of their size or the types of computers used. Many system vendors supply system administration manuals that describe the aspects of a basic program, and one can consult documents such as [FIPS73], or [NBS120].

Once a basic administration program is in place, management and users need to incorporate virus prevention measures that will help to *deter* attacks by viruses and related threats, *detect* when they occur, *contain* the attacks to limit damage, and *recover* in a reasonable amount of time without loss of data. To accomplish these aims, attention needs to be focused on the following areas:

- *educating users* about malicious software in general, the risks that it poses, how to use control measures, policies, and procedures to protect themselves and the organization

- *software management* policies and procedures that address public-domain software, and the use and maintenance of software in general

- *use of technical controls* that help to prevent and deter attacks by malicious software and unauthorized users

- *monitoring of user and software activity* to detect signs of attacks, to detect policy violations, and to monitor the overall effectiveness of policies, procedures, and controls

- *contingency policies and procedures* for containing and recovering from attacks

General guidance in each of these areas is explained in the following sections.

3.1 User Education

Education is one of the primary methods by which systems and organizations can achieve greater protection from incidents of malicious software and unauthorized use. In situations where technical controls do not provide complete protection (i.e., most computers), it is ultimately people and their willingness to adhere to security policies that will determine whether systems and organizations are protected. By educating users about the general nature of computer viruses and related threats, an organization can improve its ability to deter, detect, contain and recover from potential incidents.

Users should be educated about the following:

- how malicious software operates, methods by which it is planted and spread, the vulnerabilities exploited by malicious software and unauthorized users

- general security policies and procedures and how to use them

- the policies to follow regarding the backup, storage, and use of software, especially public-domain software and shareware

- how to use the technical controls they have at their disposal to protect themselves

- how to monitor their systems and software to detect signs of abnormal activity, what to do or whom to contact for more information

- contingency procedures for containing and recovering from potential incidents

User education, while perhaps expensive in terms of time and resources required, is ultimately a cost-effective measure for protecting against incidents of malicious software and unauthorized use. Users who are better acquainted with the destructive potential of malicious software and the methods by which it can attack systems may in turn be prompted to take measures to protect themselves. The purpose of security policies and procedures will be more clear, thus users may be more willing to actively use them. By educating users how to detect abnormal system activity and the resultant steps to follow for containing and recovering from potential incidents, organizations will save money and time if and when actual incidents occur.

3.2 Software Management

As shown by examples in Chapter 2, one of the prime methods by which malicious software is initially copied onto systems is by unsuspecting users. When users download programs from sources such as software bulletin boards, or public directories on systems or network servers, or in general use and share software that has not been obtained from a reputable source, users are in danger of spreading malicious software. To prevent users from potentially spreading malicious software, managers need to

- ensure that users understand the nature of malicious software, how it is generally spread, and the technical controls to use to protect themselves

- develop policies for the downloading and use of public-domain and shareware software

- create some mechanism for validating such software prior to allowing users to copy and use it

- minimize the exchange of executable software within an organization as much as possible

- do not create software repositories on LAN servers or in multi-user system directories unless technical controls exist to prevent users from freely uploading or downloading the software

The role of education is important, as users who do not understand the risks yet who are asked to follow necessarily restrictive policies may share and copy software anyway. Where technical controls cannot prevent placing new software onto a system, users are then primarily responsible for the success or failure of whatever policies are developed.

A policy that prohibits any copying or use of public-domain software may be overly restrictive, as some public domain programs have proved to be useful. A less restrictive policy would allow some copying, however a user might first require permission from the appropriate manager. A special system should be used from which to perform the copy and then to test the software. This type of system, called an *isolated system*, should be configured so that there is no risk of spreading a potentially malicious program to other areas of an organization. The system should not be used by other users, should not connect to networks, and should not contain any valuable data. An isolated system should also be used to test internally developed software and updates to vendor software.

Other policies for managing vendor software should be developed. These policies should control how and where software is purchased, and should govern where the software is installed and how it is to be used. The following policies and procedures are suggested:

- purchase vendor software only from reputable sources

- maintain the software properly and update it as necessary

- don't use pirated software, as it may have been modified

- keep records of where software is installed readily available for contingency purposes

- ensure that vendors can be contacted quickly if problems occur

- store the original disks or tapes from the vendor in a secure location

3.3 Technical Controls

Technical controls are the mechanisms used to protect the security and integrity of systems and associated data. The use of technical controls can help to prevent occurrences of viruses and related threats by deterring them or making it more difficult for them to gain access to systems and data. Examples of technical controls include user authentication mechanisms such as passwords, mechanisms which provide selective levels of access to files and directories (read-only, no access, access to certain users, etc.), and write-protection mechanisms on tapes and diskettes.

The different types of technical controls and the degree to which they can provide protection and deterrence varies from system to system, thus the use of specific types of controls is discussed in Chapters 4 and 5. However, the following general points are important to note:

- technical controls should be used as available to restrict system access to authorized users only

- in the multi-user environment, technical controls should be used to limit users' privileges to the minimum practical level; they should work automatically and need not be initiated by users

- users and system managers must be educated as to how and when to use technical controls

- where technical controls are weak or non-existent (i.e., personal computers), they should be supplemented with alternative physical controls or add-on control mechanisms

Managers need to determine which technical controls are available on their systems, and then the degree to which they should be used and whether additional add-on controls are necessary. One way to answer these questions is to first categorize the different classes of data being processed by a system or systems, and then to rank the categories according to criteria such as sensitivity to the organization and vulnerability of the system to attack. The rankings should then help determine the degree to which the controls should be applied and whether additional controls are necessary. Ideally, those systems with the most effective controls should be used to process the most sensitive data, and vice-versa. As an example, a personal computer which processes sensitive employee information should require add-on user authentication mechanisms, whereas a personal computer used for general word processing may not need additional controls.

It is important to note that technical controls do not generally provide complete protection against viruses and related threats. They may be cracked by determined users who are knowledgeable of hidden bugs and weaknesses, and they may be surmounted through the use of Trojan horse programs, as shown by examples in Chapter 2. An inherent weakness in technical controls is that, while deterring users and software from objects to which they do not have access, they may be totally ineffective against attacks which target objects that are accessible. For example, technical controls may not prevent an authorized user from destroying files to which the user has authorized access. Most importantly, when technical controls are not used properly, they may increase a system's degree of vulnerability. It is generally agreed that fully effective technical controls will not be widely available for some time. Because of the immediate nature of the computer virus threat, technical controls must be supplemented by less technically-oriented control measures such as described in this chapter.

3.4 General Monitoring

An important aspect of computer viruses and related threats is that they potentially can cause extensive damage within a very small amount of time, such as minutes or seconds. Through proper monitoring of software, system activity, and in some cases user activity, managers can increase their chances that they will detect early signs of malicious software and unauthorized activity. Once the presence is noted or suspected, managers can then use contingency procedures to contain the activity and recover from whatever damage has been caused. An additional benefit of general monitoring is that over time, it can aid in determining the necessary level or degree of security by indicating whether security policies, procedures, and controls are working as planned.

Monitoring is a combination of continual system and system management activity. Its effectiveness depends on cooperation between management and users. The following items are necessary for effective monitoring:

- user education - users must know, specific to their computing environment, what constitutes normal and abnormal system activity and whom to contact for further information - this is especially important for users of personal computers, which generally lack automated methods for monitoring

- automated system monitoring tools - generally on multi-user systems, to automate logging or accounting of user and software accesses to accounts, files, and other system objects - can sometimes be tuned to record only certain types of accesses such as "illegal" accesses

- anti-viral software - generally on personal computers, these tools alert users of certain types of system access that are indicative of "typical" malicious software

- system-sweep programs - programs to automatically check files for changes in size, date, or content

- network monitoring tools - as with system monitoring tools, to record network accesses or attempts to access

The statistics gained from monitoring activities should be used as input for periodic reviews of security programs. The reviews should evaluate the effectiveness of general system management, and associated security policies, procedures, and controls. The statistics will indicate the need for changes and will help to fine tune the program so that security is distributed to where it is most necessary. The reviews should also incorporate users' suggestions, and to ensure that the program is not overly restrictive, their criticisms.

3.5 Contingency Planning

The purpose of contingency planning with regard to computer viruses and related threats is to be able to contain and recover completely from actual attacks. In many ways, effective system management that includes user education, use of technical controls, software management, and monitoring activities, is a form of contingency planning, generally because a well-run, organized system or facility is better able to withstand the disruption that could result from a computer virus attack. In addition to effective system management activities, managers need to consider other contingency procedures that specifically take into account the nature of computer viruses and related threats.

Possibly the most important contingency planning activity involves the use of backups. The ability to recover from a virus attack depends upon maintaining regular, frequent backups of all system data. Each backup should be checked to ensure that the backup media has not been corrupted. Backup media could easily be corrupted because of defects, because the backup procedure was incorrect, or perhaps because the backup software itself has been attacked and modified to corrupt backups as they are made.

Contingency procedures for restoring from backups after a virus attack are equally important. Backups may contain copies of malicious software that have been hiding in the system. Restoring the malicious software to a system that has been attacked could cause a recurrence of the problem. To avoid this possibility, software should be restored only from its original media: the tapes or diskettes from the vendor. In some cases, this may involve reconfiguring the software, therefore managers must maintain copies of configuration information for system and application software. Because data is not directly executable, it can be restored from routine backups. However, data that has been damaged may need to be restored manually or from older backups. Command files such as batch procedures and files executed when systems boot or when user log on should be inspected to ensure that they have not been damaged or modified. Thus, managers will need to retain successive versions of backups, and search through them when restoring damaged data and command files.

Other contingency procedures for containing virus attacks need to be developed. The following are suggested; they are discussed in more detail in Chapters 4 and 5:

- ensure that accurate records are kept of each system's configuration, including the system's location, the software it runs, the system's network and modem connections, and the name of the system's manager or responsible individual

- create a group of skilled users to deal with virus incidents and ensure that users can quickly contact this group if they suspect signs of viral activity

- maintain a security distribution list at each site with appropriate telephone numbers of managers to contact when problems occur

- isolate critical systems from networks and other sources of infection

- place outside network connections on systems with the best protections, use central gateways to facilitate rapid disconnects

4. Virus Prevention for Multiuser Computers and Associated Networks

Virus prevention in the multi-user computer environment is aided by the centralized system and user management, and the relative richness of technical controls. Unlike personal computers, many multi-user systems possess basic controls for user authentication, for levels of access to files and directories, and for protected regions of memory. By themselves, these controls are not adequate, but combined with other policies and procedures that specifically target viruses and related threats, multi-user systems can greatly reduce their vulnerabilities to exploitation and attack.

However, some relatively powerful multi-user machines are now so compact as to be able to be located in an office or on a desk-top. These machines are still fully able to support a small user population, to connect to major networks, and to perform complex real-time operations. But due to their size and increased ease of operation, they are more vulnerable to unauthorized access. Also, multi-user machines are sometimes managed by untrained personnel who do not have adequate time to devote to proper system management and who may not possess a technical background or understanding of the system's operation. Thus, it is especially important for organizations who use or are considering machines of this nature to pay particular attention to the risks of attack by unauthorized users, viruses, and related software.

The following sections offer guidance and recommendations for improving the management and reducing the risk of attack for multi-user computers and associated networks.

4.1 General Policies

Two general policies are suggested here. They are intended for uniform adoption throughout an organization, i.e., they will not be entirely effective if they are not uniformly followed. These policies are as follows:

- An organization must assign a dedicated system manager to operate each multi-user computer. The manager should be trained, if necessary, to operate the system in a practical and secure manner. This individual should be assigned the management duties as part of his job description; the management duties should not be assigned "on top" of the individual's other duties, but rather adequate time should be taken from other duties. System management is a demanding and time-consuming operation that can unexpectedly require complete dedication. As systems are increasingly inter-connected via networks, a poorly managed system that can be used as a pathway for unauthorized access to other systems will present a significant vulnerability to an organization. Thus, the job of system manager should be assigned carefully, and adequate time be given so that the job can be performed completely.

- Management needs to impress upon users the need for their involvement and cooperation in computer security.   A method for doing this is to create an organizational security policy. This policy should be a superset of all other computer-related policy, and should serve to clearly define what is expected of the user.   It should detail how systems are to be used and what sorts of computing are permitted and not permitted. Users should read this policy and agree to it as a prerequisite to computer use. It would also be helpful to use this policy to create other policies specific to each multi-user system.

4.2 Software Management

Effective software management can help to make a system less vulnerable to attack and can make containment and recovery more successful. Carefully controlled access to software will prevent or discourage unauthorized access. If accurate records and backups are maintained, software restoral can be accomplished with a minimum of lost time and data. A policy of testing all new software, especially public-domain software, will help prevent accidental infection of a system by viruses and related software. Thus, the following policies and procedures are recommended:

- Use only licensed copies of vendor software, or software that can be verified to be free of harmful code or other destructive aspects. Maintain complete information about the software, such as the vendor address and telephone number, the license number and version, and update information. Store the software in a secure, tamper-proof location.

- Maintain configuration reports of all installed software, including the operating system. This information will be necessary if the software must be re-installed later.

- Prevent user access to system software and data. Ensure that such software is fully protected, and that appropriate monitoring is done to detect attempts at unauthorized access.

- Prohibit users from installing software. Users should first contact the system manager regarding new software. The software should then be tested on an *isolated* system to determine whether the software may contain destructive elements. The isolated system should be set up so that, to a practical degree, it replicates the target system, but does not connect to networks or process sensitive data. A highly-skilled user knowledgeable about viruses and related threats should perform the testing and ensure that the software does not change or delete other software or data. Do not allow users to directly add any software to the system, whether from public software repositories, or other systems, or their home systems.

- Teach users to protect their data from unauthorized access. Ensure that they know how to use access controls or file protection mechanisms to prevent others from reading or modifying their files. As possible, set default file protections such that when a user creates a file, the file can be accessed only by that user, and no others. Each user should not permit others to use his or her account.

- Do not set-up directories to serve as software repositories unless technical controls are used to prevent users from writing to the directory. Make sure that users contact the system manager regarding software they wish to place in a software repository. It would be helpful to track where the software is installed by setting up a process whereby users must first register their names before they can copy software from the directory.

- If developing software, control the update process so that the software is not modified without authorization. Use a software management and control application to control access to the software and to automate the logging of modifications.

- Accept system and application bug fixes or patches only from highly reliable sources, such as the software vendor. Do not accept patches from anonymous sources, such as received via a network. Test the new software on an isolated system to ensure that the software does not make an existing problem worse.

4.3 Technical Controls

Many multi-user computers contain basic built-in technical controls. These include user authentication via passwords, levels of user privilege, and file access controls. By using these basic controls effectively, managers can significantly reduce the risk of attack by preventing or deterring viruses and related threats from accessing a system.

Perhaps the most important technical control is user authentication, with the most widely form of user authentication being a username associated with a password. Every user account should use a password that is deliberately chosen so that simple attempts at password cracking cannot occur. An effective password should not consist of a person's name or a recognizable word, but rather should consist of alphanumeric characters and/or strings of words that cannot easily be guessed. The passwords should be changed at regular intervals, such as every three to six months. Some systems include or can be modified to include a password history, to prevent users from reusing old passwords. For more information on effective password practices, see [FIPS73].

The username/password mechanism can sometimes be modified to reduce opportunities for password cracking. One method is to increase the running time of the password encryption to several seconds. Another method is to cause the user login program to accept from three to five incorrect password attempts in a row before disabling the user account for several minutes. Both methods significantly increase the amount of time a password cracker would spend when making repeated attempts at guessing a password. A method for ensuring that passwords are difficult to crack involves the use of a program that could systematically guess passwords, and then send warning messages to the system manager and corresponding users if successful. The program could attempt passwords that are permutations of each user's name, as well as using words from an on-line dictionary.

Besides user authentication, access control mechanisms are perhaps the next most important technical control. Access control mechanisms permit a system manager to selectively permit or bar user access to system resources regardless of the user's level of privilege. For example, a user at a low-level of system privilege can be granted access to a resource at a higher level of privilege without raising the user's privilege through the use of an access control that specifically grants that user access. Usually, the access control can determine the type of access, e.g., read or write. Some access controls can send alarm messages to audit logs or the system manager when unsuccessful attempts are made to access resources protected by an access control.

Systems which do not use access controls usually contain another more basic form that grants access based on user categories. Usually, there are four: *owner*, where only the user who "owns" or creates the resource can access it; *group*, where anyone in the same group as the owner can access the resource; *world*, where all users can access the resource, and *system*, which supersedes all other user privileges. Usually, a file or directory can be set up to allow any combination of the four. Unlike access controls, this scheme doesn't permit access to resources on a specific user basis, thus if a user at a low level of privilege requires access to a system level resource, the user must be granted system privilege. However, if used carefully, this scheme can adequately protect users' files from being accessed without authorization. The most effective mode is to create a unique group for each user. Some systems may permit a default file permission mask to be set so that every file created would be accessible only by the file's owner.

Other technical control guidelines are as follows:

- Do not use the same password on several systems. Additionally, sets of computers that are mutually trusting in the sense that login to one constitutes login to all should be carefully controlled.

- Disable or remove old or unnecessary user accounts. Whenever users leave an organization or no longer use a system, change all passwords that the users had knowledge of.

- Practice a "least privilege" policy, whereby users are restricted to accessing resources on a need-to-know basis only. User privileges should be as restricting as possible without adversely affecting the performance of their work. To determine what level of access is required, err first by setting privileges to their most restrictive, and upgrade them as necessary. If the system uses access controls, attempt to maintain a user's system privileges at a low level while using the access controls to specifically grant access to the required resources.

- Users are generally able to determine other users' access to their files and directories, thus instruct users to carefully maintain their files and directories such that they are not accessible, or at a minimum, not writable, by other users. As possible, set default file protections such that files and directories created by each user are accessible by only that user.

- When using modems, do not provide more access to the system than is necessary. For example, if only dial-out service is required, set up the modem or telephone line so that dial-in service is not possible. If dial-in service is necessary, use modems that require an additional passwords or modems that use a call-back mechanism. These modems may work such that a caller must first identify himself to the system. If the identification has been pre-recorded with the system and therefore valid, the system then calls back at a pre-recorded telephone number.

- If file encryption mechanisms are available, make them accessible to users. Users may wish to use encryption as a further means of protecting the confidentiality of their files, especially if the system is accessible via networks or modems.

- Include software so that users can temporarily "lock" their terminals from accepting keystrokes while they are away. Use software that automatically disables a user's account if no activity occurs after a certain interval, such as 10 - 15 minutes.

4.4 Monitoring

Many multi-user systems provide a mechanism for automatically recording some aspects of user and system activity. This monitoring mechanism, if used regularly, can help to detect evidence of viruses and related threats. Early detection is of great value, because malicious software potentially can cause significant damage within a matter of minutes. Once evidence of an attack has been verified, managers can use contingency procedures to contain and recover from any resultant damage.

Effective monitoring also requires user involvement, and therefore, user education. Users must have some guidelines for what constitutes normal and abnormal system activity. They need to be aware of such items as whether files have been changed in content, date, or by access permissions, whether disk space has become suddenly full, and whether abnormal error messages occur. They need to know whom to contact to report signs of trouble and then the steps to take to contain any damage.

The following policies and procedures for effective monitoring are recommended:

- Use the system monitoring/auditing tools that are available. Follow the procedures recommended by the system vendor, or start out by enabling the full level or most detailed level of monitoring. Use tools as available to help read the logs, and determine what level of monitoring is adequate, and cut back on the level of detail as necessary. Be on the guard for excessive attempts to access accounts or other resources that are protected. Examine the log regularly, at least weekly if not more often.

- As a further aid to monitoring, use alarm mechanisms found in some access controls. These mechanisms send a message to the audit log whenever an attempt is made to access a resource protected by an access control.

- If no system monitoring is available, or if the present mechanism is unwieldy or not sufficient, investigate and purchase other monitoring tools as available. Some third-party software companies sell monitoring tools for major operating systems with capabilities that supersede those of the vendor's.

- Educate users so that they understand the normal operating aspects of the system. Ensure that they have quick access to an individual or group who can answer their questions and investigate potential virus incidents.

- Purchase or build system sweep programs to checksum files at night, and report differences from previous runs. Use a password checker to monitor whether passwords are being used effectively.

- Always report, log, and investigate security problems, even when the problems appear insignificant. Use the log as input into regular security reviews. Use the reviews as a means for evaluating the effectiveness of security policies and procedures.

- Enforce some form of sanctions against users who *consistently* violate or attempt to violate security policies and procedures. Use the audit logs as evidence, and bar the users from system use.

4.5 Contingency Planning

As stressed in Chapter 3, backups are the most important contingency planning activity. A system manager must plan for the eventuality of having to restore all software and data from backup tapes for any number of reasons, such as disk drive failure or upgrades. It has been shown that viruses and related threats could potentially and unexpectedly destroy all system information or render it useless, thus managers should pay particular attention to the effectiveness of their backup policies. Backup policies will vary from system to system, however they should be performed daily, with a minimum of several months backup history. Backup tapes should be verified to be accurate, and should be stored *off-site* in a secured location.

Viruses and related software threats could go undetected in a system for months to years, and thus could be backed up along with normal system data. If such a program would suddenly trigger and cause damage, it may require much searching through old backups to determine when the program first appeared or was infected. Therefore the safest policy is to restore programs, i.e., executable and command files, from their original vendor media only. Only system data that is non-executable should be restored from regular backups. Of course, in the case of command files or batch procedures that are developed or modified in the course of daily system activity, these may need to be inspected manually to ensure that they have not been modified or damaged.

Other recommended contingency planning activities are as follows:

- Create a security distribution list for hand-out to each user. The list should include the system manager's name and number, and other similar information for individuals who can answer users' questions about suspicious or unusual system activity. The list should indicate when to contact these individuals, and where to reach them in emergencies.

- Coordinate with other system managers, especially if their computers are connected to the same network. Ensure that all can be contacted quickly in the event of a network emergency by using some mechanism other than the network.

- Besides observing physical security for the system as well as its software and backup media, locate terminals in offices that can be locked or in other secure areas.

- If users are accessing the system via personal computers and terminal emulation software, keep a record of where the personal computers are located and their network or port address for monitoring purposes. Control carefully whether such users are uploading software to the system.

- Exercise caution when accepting system patches. Do not accept patches that arrive over a network unless there is a high degree of certainty as to their validity. It is best to accept patches only from the appropriate software vendor.

4.6 Associated Network Concerns

Multi-user computers are more often associated with relatively large networks than very localized local area networks or personal computer networks that may use dedicated network servers. The viewpoint taken here is that wide area network and large local area network security is essentially a collective function of the systems connected to the network, i.e., it is not practical for a controlling system to monitor *all* network traffic and differentiate between authorized and unauthorized use. A system manager should generally assume that network connections pose inherent risks of unauthorized access to the system in the forms of unauthorized users and malicious software. Thus, a system manager needs to protect the system from network-borne threats and likewise exercise responsibility by ensuring that his system is not a source of such threats, while at the same time making network connections available to users as necessary. The accomplishment of these aims will require the use of technical controls to restrict certain types of access, monitoring to detect violations, and a certain amount of trust that users will use the controls and follow the policies.

Some guidelines for using networks in a more secure manner are as follows:

- Assume that network connections elevate the risk of unauthorized access. Place network connections on system which provide adequate controls, such as strong user authentication and access control mechanisms. Avoid placing network connections on system which process sensitive data.

- If the system permits, require an additional password or form of authentication for accounts accessed from network ports. If possible, do not permit access to system manager accounts from network ports.

- If anonymous or guest accounts are used, place restrictions on the types of commands that can be executed from the account. Don't permit access to software tools, commands that can increase privileges, and so forth.

- As possible, monitor usage of the network. Check if network connections are made at odd hours, such as during the night, or if repeated attempts are made to log in to the system from a network port.

- When more than one computer is connected to the same network, arrange the connections so that one machine serves as a central gateway for the other machines. This will allow a rapid disconnect from the network in case of an attack.

- Ensure that users are fully educated in network usage. Make them aware of the additional risks involved in network access. Instruct them to be on the alert for any signs of tampering, and to contact an appropriate person if they detect any suspicious activity. Create a policy for responsible network usage that details what sort of computing activity will and will not be tolerated. Have users read the policy as a prerequisite to network use.

- Warn users to be suspicious of any messages that are received from unidentified or unknown sources.

- Don't advertise a system to network users by printing more information than necessary on a welcome banner. For example, don't include messages such as "Welcome to the Payroll Accounting System" that may cause the system to be more attractive to unauthorized users.

- Don't network to outside organizations without a mutual review of security practices

5. Virus Prevention for Personal Computers and Associated Networks

Virus prevention in the personal computer environment differs from that of the multi-user computer environment mainly in the following two respects: the relative lack of technical controls, and the resultant emphasis this places on less-technically oriented means of protection which necessitates more reliance on user involvement. Personal computers typically do not provide technical controls for such things as user authorization, access controls, or memory protection that differentiates between system memory and memory used by user applications. Because of the lack of controls and the resultant freedom with which users can share and modify software, personal computers are more prone to attack by viruses, unauthorized users, and related threats.

Virus prevention in the personal computer environment must rely on continual user awareness to adequately detect potential threats and then to contain and recover from the damage. Personal computer users are in essence personal computer managers, and must practice their management as a part of their general computing. Personal computers generally do not contain auditing features, thus a user needs to be aware at all times of the computer's performance, i.e., what it is doing, or what is normal or abnormal activity. Ultimately, personal computer users need to understand some of the technical aspects of their computers in order to protect, deter, contain, and recover. Not all personal computer users are technically oriented, thus this poses some problems and places even more emphasis on user education and involvement in virus prevention.

Because of the dependance on user involvement, policies for the personal computer environment are more difficult to implement than in the multi-user computer environment. However, emphasizing these policies as part of a user education program will help to ingrain them in users' behavior. Users should be shown via examples what can happen if they don't follow the policies. An example where users share infected software and then spread the software throughout an organization would serve to effectively illustrate the point, thus making the purpose of the policy more clear and more likely to be followed. Another effective method for increasing user cooperation is to create a list of effective personal computer management practices specific to each personal computing environment. Creating such a list would save users the problem of determining how best to enact the policies, and would serve as a convenient checklist that users could reference as necessary.

It will likely be years before personal computers incorporate strong technical controls in their architectures. In the meantime, managers and users must be actively involved in protecting their computers from viruses and related threats.   The following sections provide guidance to help achieve that aim.

5.1   General Policies

Two general policies are suggested here.   The first requires that management make firm, unambiguous decisions as to how users should operate personal computers, and state that policy in writing.   This policy will be a general re-statement of all other policies affecting personal computer use.   It is important that users read this policy and agree to its conditions as a prerequisite to personal computer use.   The purposes of the policy are to   (1) ensure that users are aware of all policies, and (2) impress upon users the need for their active involvement in computer security.

The second policy is that every personal computer should have an "owner" or "system manager" who is responsible for the maintenance and security of the computer, and for following all policies and procedures associated with the use of the computer.   It would be preferable that the primary user of the computer fill this role.   It would not be too extreme to make this responsibility a part of the user's job description.   This policy will require that resources be spent on educating users so that they can adequately follow all policies and procedures.

5.2   Software Management

Due to the wide variety of software available for many types of personal computers, it is especially important that software be carefully controlled.   The following policies are suggested:

- Use only licensed copies of vendor software for personal computers.   Ensure that the license numbers are logged, that warranty information is completed, and that updates or update notices will be mailed to the appropriate users.   Ensure that software versions are uniform on all personal computers.   Purchase software from known, reputable sources - do not purchase software that is priced suspiciously low and do not use pirated software, even on a trial basis.   As possible, buy software with built-in security features.

- Do not install software that is not clearly needed.   For example, software tools such as compilers or debuggers should not be installed on machines where they are not needed.

- Store the original copies of vendor software in a secure location for use when restoring the software.

- Develop a clear policy for use of public-domain software and shareware. It is recommended that the policy prohibit indiscriminate downloading from software bulletin boards. A special *isolated* system should be configured to perform the downloading, as well as for testing downloaded and other software or shareware. The operation of the system should be managed by a technically skilled user who can use anti-virus software and other techniques to test new software before it is released for use by other users.

- Maintain an easily-updated database of installed software. For each type of software, the database should list the computers where the software is installed, the license numbers, software version number, the vendor contact information, and the responsible person for each computer listed. This database should be used to quickly identify users, machines, and software when problems or emergencies arise, such as when a particular type of software is discovered to contain a virus or other harmful aspects.

- Minimize software sharing within the organization. Do not permit software to be placed on computers unless the proper manager is notified and the software database is updated. If computer networks permit software to be mailed or otherwise transferred among machines, prohibit this as a policy. Instruct users not to run software that has been mailed to them.

- If using software repositories on LAN servers, set up the server directory such that users can copy from the directory, but not add software to the directory. Assign a user to manage the repository; all updates to the repository should be cleared through this individual. The software should be tested on an isolated system as described earlier.

- If developing software, consider the use of software management and control programs that automate record keeping for software updates, and that provide a degree of protection against unauthorized modifications to the software under development.

- Prohibit users from using software or disks from their home systems. A home system that is used to access software bulletin boards or that uses shared copies of software could be infected with viruses or other malicious software.

5.3 Technical Controls

As stated earlier, personal computers suffer from a relative lack of technical controls. There are usually no mechanisms for user authentication and for preventing users or software from modifying system and application software. Generally, all software and hardware is accessible by the personal computer user, thus the potential for misuse is substantially greater than in the multi-user computer environment.

However, some technical controls can be added to personal computers, e.g., user authentication devices. The technical controls that do not exist can be simulated by other controls, such as a lock on an office door to substitute for a user authentication device, or anti-virus software to take the place of system auditing software. Lastly, some of the personal computer's accessibility can be reduced, such as by the removal of floppy diskette drives or by the use of diskless computers that must download their software from a LAN server. The following items are suggested:

- Where technical controls exist, use them. If basic file access controls are available to make files read-only, make sure that operating system files and other executable files are marked as read-only. Use write-protect tabs on floppy diskettes and tapes. If LAN access requires a password, ensure that passwords are used carefully - follow the guidelines for password usage presented in Chapter 4 or see [FIPS73].

- Use new cost-effective forms of user identification such as magnetic access cards. Or, setup other software such as password mechanism that at a minimum deters unauthorized users.

- If using a LAN, consider downloading the personal computer's operating system and other applications from a read-only directory on the LAN server (instead of the personal computer's hard disk). If the LAN server is well protected, this arrangement would significantly reduce chances of the software becoming infected, and would simplify software management.

- Consider booting personal computers from write-protected floppy diskettes (instead of the computer's hard disk). Use a unique diskette per computer, and keep the diskette secured when not in use.

- Do not leave a personal computer running but unattended. Lock the computer with a hardware lock (if possible), or purchase vendor add-on software to "lock" the keyboard using a password mechanism. Alternatively turn off the computer and lock the office door. Shut down and lock the computer at the end of the day.

- When using modems connected to personal computers, do not provide more access to the computer than necessary. If only dial-out service is required, configure the modem so that it won't answer calls. If dial-in service is necessary, consider purchasing modems that require a password or that use a call-back mechanism to force a caller to call from a telephone number that is known to the modem.

- Consider using "limited-use" systems, whereby the capabilities of a system are restricted to only what is absolutely required. For example, users who run only a certain application (such as word-processor) may not require the flexibility of a personal computer. At the minimum, do not install applications or network connections where they are not needed.

5.4 Monitoring

Personal computer operating systems typically do not provide any software or user monitoring/auditing features. Monitoring, then, is largely a user function whereby the user must be aware of what the computer is doing, such as when the computer is accessing the disk or the general speed of its response to commands, and then must decide whether the activity is normal or abnormal. Anti-viral software can be added to the operating system and run in such a way that the software flags or in some way alerts a user when suspicious activity occurs, such as when critical files or memory regions are written.

Effective monitoring depends on user education. Users must know what constitutes normal and abnormal activity on their personal computers. They need to have a reporting structure available so that they can alert an informed individual to determine whether there is indeed a problem. They need to know the steps to take to contain the damage, and how to recover. Thus, the following policies and procedures are recommended:

- Form a team of skilled technical people to investigate problems reported by users. This same group could be responsible for other aspects of virus prevention, such as testing new software and handling the containment and recovery from virus-related incidents. Ensure that users have quick access to this group, e.g., via a telephone number.

- Educate users so that they are familiar with how their computers function. Show them how to use such items as anti-viral software. Acquaint them with how their computers boot, what files are loaded, whether start-up batch files are executed, and so forth.

- Users need to watch for changes in patterns of system activity. They need to watch for program loads that suddenly take longer, whether disk accesses seem excessive for simple tasks, do unusual error messages occur, do access lights for disks turn on when no disk activity should occur, is less memory available than usual, do files disappear mysteriously, is there less disk space than normal?

- Users also need to examine whether important files have changed in size, date, or content. Such files would include the operating system, regularly-run applications, and other batch files. System sweep programs may be purchased or built to perform checksums on selected files, and then to report whether changes have occurred since the last time the program was run.

- Purchase virus prevention software as applicable. At a minimum, use anti-viral software to test new software before releasing it to other users. However, do not download or use pirated copies of anti-viral software.

- Always report, log, and investigate security problems, even when the problems appear insignificant. Then use the log as input into regular security reviews. Use the reviews as a means for evaluating the effectiveness of security policies and procedures.

5.5 Contingency Planning

As described in Chapter 3, backups are the single most important contingency procedure. It is especially important to emphasize regular backups for personal computers, due to their greater susceptibility to misuse and due to the usual requirement of direct user involvement in the backup procedure, unlike that of multi-user computers. Because of the second factor, where users must directly copy files to one or more floppy diskettes, personal computer backups are sometimes ignored or not done completely. To help ensure that backups are done regularly, external backup mechanisms that use a high-density tape cartridge can be purchased and a user assigned to run the backup procedure on a regular basis. Additionally, some personal computer networks contain a personal computer backup feature, where a computer can directly access a network server's backup mechanism, sometimes in an off-line mode at a selected time. If neither of these mechanisms are available, then users must be supplied with an adequate number of diskettes to make complete backups and to maintain a reasonable amount of backup history, with a minimum of several weeks.

Users should maintain the original installation media for software applications and store it in a secure area, such as a locked cabinet, container, or desk. If a user needs to restore software, the user should use only the original media; the user should not use any other type of backup or a copy belonging to another user, as they could be infected or damaged by some form of malicious software.

The effectiveness of a backup policy can be judged by whether a user is able to recover with a minimum loss of data from a situation whereby the user would have to format the computer's disk and reload all software. Several incidents of malicious software have required that users go to this length to recover - see [MACAFEE89].

Other important contingency procedures are described below:

- Maintain a database of personal computer information. Each record should include items such as the computer's configuration, i.e., network connections, disks, modems, etc., the computer's location, how it is used, the software it runs, and the name of the computer's primary user/manager. Maintain this database to facilitate rapid communication and identification when security problems arise.

- Create a security distribution list for each user. The list should include names of people to contact who can help identify the cause of unusual computer activity, and other appropriate security personnel to contact when actual problems arise.

- Create a group of skilled users who can respond to users' inquiries regarding virus detection. This group should be able to determine when a computer has been attacked, and how best to contain and recover from the problem.

- Set up some means of distributing information rapidly to all affected users in the event of an emergency. This should not rely upon a computer network, as the network could actually be attacked, but could use other means such as telephone mail or a general announcement mechanism.

- Observe physical security for personal computers. Locate them in offices that can be locked. Do not store software and backups in unsecured cabinets.

5.6 Associated Network Concerns

Personal computer networks offer many advantages to users, however they must be managed carefully so that they do not increase vulnerability to viruses and related threats. Used incorrectly, they can become an additional pathway to unauthorized access to systems, and can be used to plant malicious software such as network worms. This section does not provide specific management guidance, as there are many different types of personal computer networks with widely varying degrees of similarity. However, some general suggestions for improving basic management are listed below:

- Assign a network administrator, and make the required duties part of the administrator's job description. Personal computer networks are becoming increasingly complex to administer, thus the administration should not be left to an individual who cannot dedicate time as necessary.

- Protect the network server(s) by locating them in secure areas. Make sure that physical access is restricted during off-hours. If possible, lock or remove a server's keyboard to prevent tampering.

- Do not provide for more than one administrator account, i.e., do not give other users administrator privileges. Similar to the problem of multiple system manager accounts on multi-user systems, this situation makes it more likely that a password will become known, and makes overall management more difficult to control. Users should coordinate their requests through a single network administrator.

- Do not permit users to connect personal computers to the network cable without permission. The administrator should keep an updated diagram of the network's topology, complete with corresponding network addresses and users.

- Use the network monitoring tools that are available. Track network usage and access to resources, and pinpoint unauthorized access attempts. Take appropriate action when violations consistently occur, such as requiring the user in question to attend a network user class or disabling the user's network account.

- Ensure that users know how to properly use the network. Show them how to use all security features. Ensure that users know how to use passwords and access controls effectively - see [FIPS73] for information on password usage. Show them the difference between normal and abnormal network activity or response. Encourage users to contact the administrator if they detect unusual activity. Log and investigate all problems.

- Do not give users more access to network resources than they require. If using shared directories, make them read-only if write permission is not required, or use a password. Encourage users to do the same with their shared directories.

- Do not set up directories for software repository unless (1) someone can first verify whether the software is not infected, and (2) users are not permitted to write to the directory without prior approval.

- Backup the network server(s) regularly. If possible or practical, backup personal computers using the network server backup mechanism.

- Disable the network mail facility from transferring executable files, if possible. This will prevent software from being indiscriminately shared, and may prevent network worm programs from accessing personal computers.

- For network guest or anonymous accounts, limit the types of commands that can be executed.

- Warn network users to be suspicious of any messages or programs that are received from unidentified sources - network users should have a critical and suspicious attitude towards anything received from an unknown source.

- Always remove old accounts or change passwords. Change important passwords immediately when users leave the organization or no longer require access to the network.

Appendix A—References

BUNZEL88   Bunzel, Rick; Flu Season; Connect, Summer 1988.

DENNING88   Denning, Peter J.; Computer Viruses; American Scientist, Vol 76, May-June, 1988.

DENNING89   Denning, Peter J.; The Internet Worm; American Scientist, Vol 77, March-April, 1989.

FIPS73   Federal Information Processing Standards Publication 73, Guidelines for Security of Computer Applications; National Bureau of Standards, June, 1980.

FIPS112   Federal Information Processing Standards Publication 112, Password Usage; National Bureau of Standards, May, 1985.

MACAFEE89   McAfee, John; The Virus Cure; Datamation, Feb 15, 1989.

NBS120   NBS Special Publication 500-120; Security of Personal Computer Systems: A Management Guide; National Bureau of Standards, Jan 1985.

SPAFFORD88   Spafford, Eugene H.; The Internet Worm Program: An Analysis; Purdue Technical Report CSD-TR-823, Nov 28, 1988.

THOMPSON84 Thompson, Ken; Reflections on Trusting Trust (Deliberate Software Bugs); Communications of the ACM, Vol 27, Aug 1984.

Appendix B—Suggested Reading

In addition to the references listed in Appendix A, the following documents are suggested reading for specific and general information on computer viruses and related forms, and other related security information.

Brenner, Aaron; LAN Security; LAN Magazine, Aug 1989.

Cohen, Fred; Computer Viruses, Theory and Experiments; 7th Security Conference, DOD/NBS Sept 1984.

Computer Viruses - Proceedings of an Invitational Symposium, Oct 10/11, 1988; Deloitte, Haskins, and Sells; 1989

Dvorak, John; Virus Wars: A Serious Warning; PC Magazine; Feb 29, 1988.

Federal Information Processing Standards Publication 83, Guideline on User Authentication Techniques for Computer Network Access Control; National Bureau of Standards, Sept, 1980.

Federal Information Processing Standards Publication 87, Guidelines for ADP Contingency Planning; National Bureau of Standards, March, 1981.

Fiedler, David and Hunter, Bruce M.; Unix System Administration; Hayden Books, 1987

Fitzgerald, Jerry; Business Data Communications: Basic Concepts, Security, and Design; John Wiley and Sons, Inc., 1984

Gasser, Morrie; Building a Secure Computer System; Van Nostrand Reinhold, New York, 1988.

Grampp, F. T. and Morris, R. H.; UNIX Operating System Security; AT&T Bell Laboratories Technical Journal, Oct 1984.

Highland, Harold J.; From the Editor -- Computer Viruses; Computers & Security; Aug 1987.

Longley, Dennis and Shain, Michael; Data and Computer Security

NBS Special Publication 500-120; Security of Personal Computer Systems: A Management Guide; National Bureau of Standards, Jan 1985.

Parker, T.; Public domain software review: Trojans revisited, CROBOTS, and ATC; Computer Language; April 1987.

Schnaidt, Patricia; Fasten Your Safety Belt; LAN Magazine, Oct 1987.

Shoch, J. F. and Hupp, J. A.; The Worm Programs: Early Experience with a Distributed Computation; Comm of ACM, Mar 1982.

White, Stephen and Chess, David; Coping with Computer Viruses and Related Problems; IBM Research Report RC 14405 (#64367), Jan 1989.

Witten, I. H.; Computer (In)security: infiltrating open systems; Abacus (USA) Summer 1987.

Printed and bound by CPI Group (UK) Ltd, Croydon, CR0 4YY

03/10/2024

01040333-0009